WOODWORKING TOOLS

CHRISTIE'S COLLECTORS GUIDES

WOODWORKING TOOLS

CHRISTOPHER PROUDFOOT
AND
PHILIP WALKER

PHAIDON · CHRISTIE'S
OXFORD

© Phaidon · Christie's Limited 1984

First published in Great Britain in 1984 by
Phaidon · Christie's Ltd., Littlegate House, St Ebbe's Street, Oxford OX1 1SQ

British Library Cataloguing in Publication Data

Proudfoot, Christopher
   Woodworking tools
      1. Woodworking tools—Collectors and collecting
      I. Title   II. Walker, Philip
      621.9'075   TT186

      ISBN 0–7148–8005–1

Printed and bound in Spain by Heraclio Fournier S.A., Vitoria

Frontispiece: Painted coachbuilder's chest. Tool-chests often have beautifully inlaid interiors (the exteriors are always very plain), but a painted interior of this standard is much less common. The chest and its contents belonged to one John Hartley, who was apprenticed in 1838. The lid design incorporates the Royal Arms as well as those of the Company of Coachmakers and Coach-Harness-Makers of London, with their crest of Phoebus driving the chariot of the Sun. The tools visible in the picture are various coachbuilders' routers and T-rebate planes.
(*Old Woodworking Tools*)

CONTENTS

ACKNOWLEDGEMENTS

A particular joy of writing this book has been the help offered by collectors and dealers. Mention must first be made of the part played by Tony Barwick, whose Islington establishment Old Woodworking Tools provided material for many of the illustrations, and even offered on-site facilities for the photography. Others who lent tools for photography included Richard Maude and Bill Rhea. My co-author Philip Walker not only contributed chapter four but also checked the text of most of the other chapters. Tom Seward did the same for chapter three and offered some useful information for it. Roy Arnold provided the bibliography; his stock of books on tools made him uniquely qualified to do this.

I would like to thank Gordon Roberton and Ted Holmes, two hard-worked photographers who have slaved over hot cameras to produce illustrations, and finally I must thank those who have made the book possible by publishing information on tools which I would never have had time to research myself. Their names appear in the text and in the bibliography, but I would mention in particular W.L. Goodman, R.A. Salaman, Ken Roberts, Alvin Sellens and Roger K. Smith.

Where dates are given for plane-makers and plane-iron makers, these are the dates given by Goodman in *British Planemakers from 1700*.

Captions followed by the letter (P) indicate that the objects illustrated are listed in the Price Guide on page 154.

INTRODUCTION

In an age when almost anything, from bricks to barbed wire, is collected by someone, it will come as no surprise that tools are a well established collecting subject, particularly those of the various woodworking trades.

The most familiar of these trades are those of the carpenter (whose tools may consist of little more than hammer, saw, adze or jack plane and try-square) the joiner and the cabinet-maker. The latter two are closely allied, the former concentrating more on architectural joinery and the latter on furniture – particularly, furniture which is veneered in fine woods. There are very few tools peculiar to the carpenter or the cabinet-maker, the great majority being shared with the joiner, but there are specialist trades which have tools unused by any other: notably, coach-building, wheelwrighting and coopering.

By far the greatest scope in tool collecting occurs in the field of planes; apart from the smoothing and jack or 'bench' planes, which are the type most familiar to the layman, there was in the past a vast range of special-purpose planes for grooving and shaping, most of which are now obsolete, partly because of the lack of ornament in modern design and partly through the use of machinery. The first chapter covers most of the traditional wooden planes used in Great Britain (which were also adopted in North America) and the second chapter concentrates on a peculiarly British phenomenon, the high-quality metal plane developed in the eighteenth and nineteenth centuries, and particularly suited to the precision joinery found in furniture of that period. These planes continued to be made up to and even after the Second World War, but by then both they and the traditional wooden patterns were being replaced by the efficient, mass-produced iron planes developed in America in the second half of the nineteenth century.

Chapter three deals exclusively with these American planes, which were produced in sufficient variety of design and ingenuity to form a subject for collecting on their own. Many are already obsolete and even while this book was in preparation, Record of Sheffield announced the deletion from their catalogues of two important planes in this tradition.

Chapter four is contributed by Philip Walker, a well known and leading authority on old tools for many years, and takes a look at some of the more fanciful planes and other tools, many of them carved with their date of manufacture, that were made mainly on the continent of Europe rather than in the English-speaking world.

Chapter five looks at hole-boring tools, particularly braces; here again, it was in Great Britain in the nineteenth century that the traditional, simple wooden brace was developed into a high-quality brass-framed or brass-reinforced implement and again it was the Americans who developed a simple, efficient metal replacement that is now to be seen in every workshop.

The final chapter guides the reader briefly through some of the remaining tools, including chisels, axes and adzes, hammers and tools used in measuring and marking out timber. Although some of these are very basic parts of a tool-kit, they can be surprisingly difficult to find in early or interesting forms: this applies particularly to hammers, chisels and saws. It is perhaps for this reason that interest has been concentrated so much on planes.

What may surprise many newcomers to tool-collecting is the amount of literature already available on the subject. However, very little of this will be found in ordinary bookshops, and apart from Goodman's *History of Woodworking Tools* (first published in 1964, but still essential reading for basic background information) most is aimed at the established collector who is anxious for more detailed information than would appeal to the beginner. It is this imbalance that I have tried to correct in this book by distilling into an easily assimilated form some of the information to be found in the more learned works of such authors as Goodman, Salaman, Sellens, Roberts and Smith.

It is my hope that I may introduce many new enthusiasts to the hobby of tool-collecting, who will in turn come to regard these heavier tomes as regular bedtime reading.

Christopher Proudfoot

1 WOOD PLANES

Wood was the normal material for plane construction for centuries and is still used in many parts of the world. In Great Britain, many High Street toolshops in the early 1960s still had beechwood planes in their windows. Although various woods were used by continental makers, and joiners who made their own planes would pick any suitably sized offcut that came to hand, the British plane-makers invariably chose beech. Beech is hard, resistant to splitting and very stable – that is, it is less liable to warp than elm, for example. It has little resistance to rot, but this is only a problem with planes that are no longer in use and are stored in damp conditions.

The planes most familiar to newcomers to this subject will be bench planes. This is the term normally given to the basic planes of any woodworker's kit – smoothing, jack and trying planes. The first of these is normally about eight inches in length and coffin-shaped; jack planes are straight-sided and usually fourteen to eighteen inches long and trying planes are similar but range in size from twenty-two inches to twenty-eight (longer planes are known as jointers). Jack and trying planes have a handle to the rear of the cutting iron, tradionally of open form (and easily split) on a jack and closed, like a saw-handle, on a trying plane. Until early in the nineteenth century, these handles were usually offset slightly to the right (the side furthest from a right-handed user). The theory behind this, presumably, was that it would counteract any tendency of the plane to steer away from a user standing in front of the bench, and keep it on a straight course. In practice, one does not always want to keep the plane parallel to the sides of the timber, and the centrally placed handle seems more comfortable.

The shape of English bench planes hardly changed otherwise in the eighteenth and nineteenth centuries, and an early example could easily be overlooked in a box of later planes. Early bench planes are, in any case, quite rare; nearly all the early planes by known makers that are found are moulding planes of one kind or another. The reason for this is that bench planes are subject to everyday use, and

1
Basic wood planes (from the 1925 catalogue of Richard Melhuish and Co. of London).

2
Fielding planes: all these date from the 18th century, that in the centre being the earliest: it is dated 1745, and carries the initials of Robert Bottle, of the Bottle family of joiners from Harrietsham, in Kent. The plane on the left is by Fitkin, that on the right by John Green of York. On both the larger planes, the handle is well offset in typical 18th-century fashion.
(*Old Woodworking Tools*)

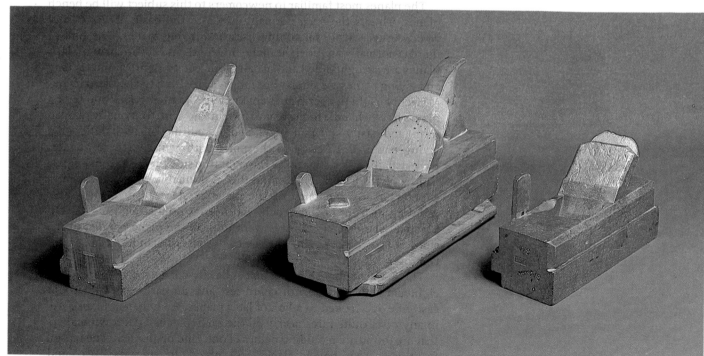

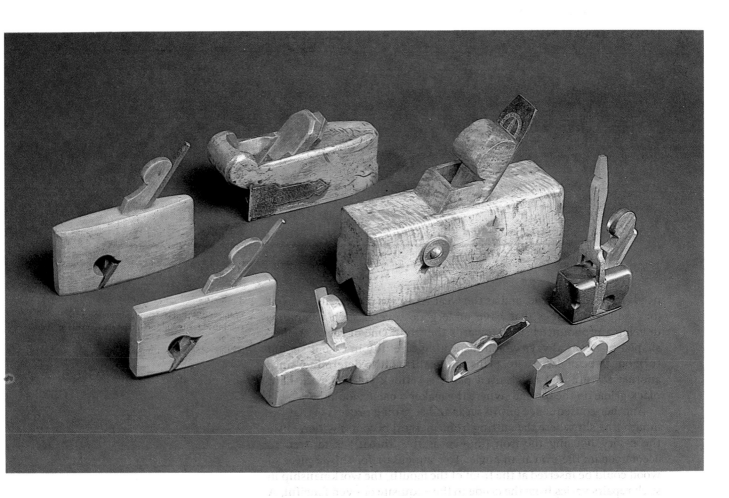

3

User-made planes: *left to right, top to bottom*: two thumb rebate planes in boxwood (each 4 in. long); boxwood bullnose plane with iron-plated toe; boxwood router; maplewood stop chamfer plane (6¼ in. long); two small bronze bullnose shoulder planes; dowel rounder, for mounting in a brace (used for rounding the locating dowels used in piano-making, for example).
(*Old Woodworking Tools*)

inevitably wear out eventually. Even planes of early appearance that do turn up tend to lack a maker's name, and it may be that many bench planes in the eighteenth century were made by their users, even after it became normal to buy in ready-made moulding planes.

Certainly, many owners repaired or modified their planes; handles were sometimes added to smoothers, and various methods were adopted for compensating for wear on the sole. Many smoothers are found with a cast-iron 'shoe' on the foward part of the sole, or sometimes the entire sole, held in place by a screw passing through from the top of the stock. A less satisfactory method was to screw a plate of brass, steel or even aluminium on to the sole with ordinary wood-screws. Even when the latter are properly countersunk, this method does not contribute to a fine finish. The purpose-made iron shoes added considerably to the weight of the plane (useful in dealing with awkward hardwoods), and were sometimes added by the makers.

Iron soles are seldom, if ever, found on jack or larger bench planes; the extra weight would have been a serious disadvantage, and in any case, wear is less rapid as the sole is larger. When the sole does wear, it

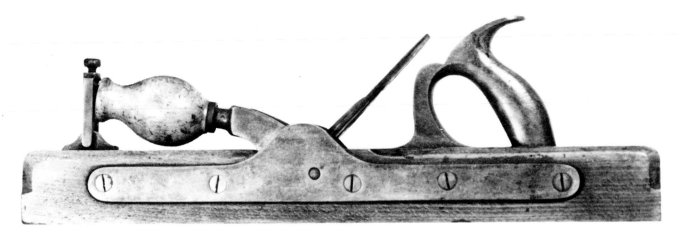

needs to be planed true, and as the wear is always greater at the front (or 'toe') end, planes that have undergone several such trimmings develop a pronounced slope towards the front. One of these 'wedge' shaped planes is always worth inspecting carefully in case it should turn out to be of early date. (This can be difficult to determine, since another feature of a much-used plane is a thick layer of rock-hard black grime on the toe end, where the maker's name would be.)

Another consequence of this wear and re-truing process is that the mouth (the slit where the cutting iron emerges) becomes wider, since the cavity housing the iron (known as the 'throat') is tapered to accommodate the iron at an angle. To compensate for this, a piece of wood could be inserted at the front of the mouth; the workmanship in such repairs varies from the crude to the exquisite or even fanciful. A further form of wear, for which no repair is necessary, is a depression on the side of the stock, where the thumb of the user's left hand normally rests. I have seen such depressions up to half-an-inch deep, suggesting that a previous owner had a vice-like grip and skin like sandpaper.

4
Crane and Windlass try-plane: an ingenious combined lever and fore-grip takes the place of the wedge.
(Christie's South Kensington)   (P)

5
Mouth repair: an unusually imaginative way of compensating for the wear and widening of the mouth, using boxwood and ebony.
(Richard Maude Collection)
(Christie's South Kensington photo)

Patina is an important aspect to look for when considering the age of a plane; it is hard to describe, but the black encrustation mentioned above is part of it, and contrasts vividly with the polished, brownish-red parts that are constantly handled when the plane is in use. The rather insipid pinkish-fawn colour of raw beech mellows with age, use and constant applications of linseed oil, and often achieves a colour that is hard to distinguish from mahogany. Although a good patina is a very attractive feature, however, it is not always a sign of great age, any more than a plane looking almost new and unused is necessarily of recent manufacture.

Many planes that look at first sight to be in little-used condition have in fact had the misfortune to be cleaned, often with all the old surface removed, and even treated with a coat of varnish. While one hopes that few collectors would commit such vandalism, such treatment occurs all too often when a householder finds Uncle Joe's old tool-chest in the garden shed and decides to clean up the contents before selling them, little realizing that he is reducing their desirability to most buyers with every stroke of his sanding block. Traces of the original dark colour may remain in the more inaccessible angles, and a single coat of varnish, particularly slow-drying varnish such as the ubiquitous polyurethane, usually leaves a hard, rough surface with very little gloss.

The blade in a plane is seldom called a blade, but is known rather as the 'iron' or 'cutting iron'; the term 'cutter' is also used, but normally only in referring to metal planes. From the end of the eighteenth century, if not earlier, it became normal for bench planes, at least in Great Britain, to be fitted with double irons. These had a slightly bowed steel plate known as a 'cap iron' or 'back iron' clamped, usually by a screw, to the iron proper, with the lower end of the cap iron slightly behind the cutting edge. The purpose of the double iron was (and is) to make the shaving curl away from the iron, so that it is less likely that the grain of the wood in front of the cutting edge will tear up. The shaving will also clear itself from the cutter more easily. On the very earliest examples, the cap iron was apparently held in place only by the wedge, which meant that some dexterity would be required in setting both the iron and cap iron while tightening the wedge, but most examples have a short screw passing through a long slot in the iron into a threaded hole in the cap iron. Most cap irons seen have a domed brass nut riveted in place over this hole to provide more 'bite' for the screw, but a recessed square nut may be found on some very early irons, and some early nineteenth-century ones have no nut at all — merely a hole punched in the metal of the cap iron with an undersized punch, so as to leave a greater depth of metal for the thread when the hole is tapped out to the correct size.

Early irons have rounded tops, with cap irons to match, but in the early years of the nineteenth century this style gave way to the more familiar bevelled shape, which lends itself more to adjusting with a

hammer. The shape has survived into the present era of lever and screw adjustment, although now the sharp corners are sometimes ground off.

On wooden planes, the irons were normally tapered in thickness, with the thickest part at the cutting edge. Only the lower part (below the cap-iron screw slot) was normally of steel, and that was a sort of thick veneer, fire-welded on to a wrought iron base. On moulding plane irons, the steel was often in the region of $\frac{1}{16}$-inch thick only. On a newly ground iron, the change from bright steel to grey iron is usually visible, and the difference between the two metals is instantly noticeable if, as is often necessary on moulding plane irons, re-shaping is done with a file. One reason for this method of construction was no doubt the great expense of tool steel, but it is also a sound practice, in that it enables very hard steel to be used for the cutting edge, with less brittle iron forming the body of the cutter.

Plane irons were normally made not by the makers of the planes

6

User-made planes: *left to right, top to bottom*: double bead plane (left- and right-handed, for use on awkwardly grained timber so that the plane can be worked in either direction); thumb bullnose plane in ebony with bone or ivory wedge; spill plane (held in a vice while scrap wood was pulled along the cutting edge to produce lighting spills); two sill planes, in mahogany and beech, for forming the 'throating' on the shoulders on window-sills; mahogany router; steel-soled left and right hand side-rebate plane.
(*Old Woodworking Tools*)

themselves, but by specialist makers who often also made edge tools such as chisels. There is seldom, if ever, any reason for assuming that the name on a plane iron is that of the maker of the plane, even when that maker is known to have been a plane-maker. Irons tend to be replaced over the years as they wear out, and, in addition, may often get switched from plane to plane in a workshop where there is more than one plane of similar width. Moulding plane irons are, of course, much less likely to be non-original, as they receive much less wear and are not interchangeable. Early plane-iron makers to look out for include Robert Moore (1750–70) and James Cam (1781 or earlier – 1838). Although large firms like Marples in the present century made their own irons, earlier plane-makers' names appearing on irons probably indicate that the irons were made under contract by a specialist. All the same, it is pleasing to find an early plane with an iron bearing the same maker's name, and also to find a plane of any

7
Contrast in colours: at the top is a badger plane in yew-wood, an unusual timber for plane-making. Below, on the left is a side fillister that has been over-enthusiastically cleaned, leaving a blotchy appearance with dark specks where old dents have retained the dirt. On the right is a moulding plane by William Moss (Birmingham, 1775–1843) which appears never to have been used, the beech hardly having darkened at all. The moulding is unusual: a Grecian ogee without a quirk.
(*Old Woodworking Tools*)

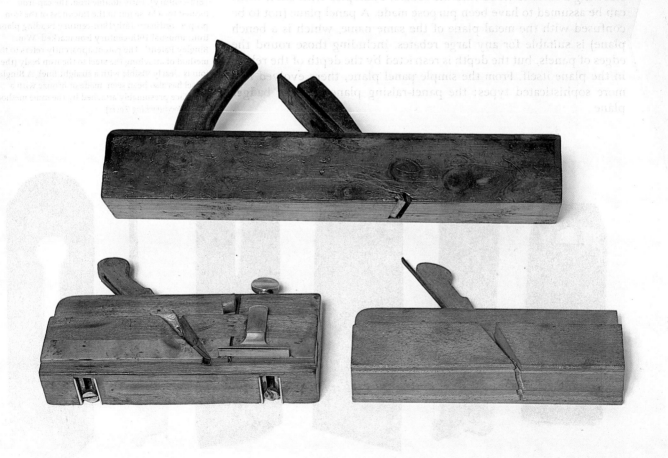

date with an iron by an early specialist maker such as Robert Moore. If plane irons of this period are rare, it is probably true to say that they at least have a higher survival rate than chisels by the same makers.

Special Purpose Planes

From bench planes it seems logical to move on to those special purpose planes whose design derives from bench planes; most of these are used in forming rebates of one kind or another.

The panel plane is similar to a jack plane in shape and, usually, size, although some examples are hardly longer than a smoothing plane. Along the lower right-hand edge of the stock is a rebate which allows the cutter to work into a rebate, and some such planes appear to be jack planes which have been modified by their owners; whether modified or not, such a plane can always be used as a jack plane by screwing a fillet of wood into the rebate. Examples with skew irons can be assumed to have been purpose made. A panel plane (not to be confused with the metal plane of the same name, which is a bench plane) is suitable for any large rebates, including those round the edges of panels, but the depth is restricted by the depth of the rebate in the plane itself. From the simple panel plane, there evolved two more sophisticated types: the panel-raising plane and the badger plane.

9

Plough and other special purpose planes (Melhuish, 1925). The Melhuish's plough and the jack plane shown next to it were probably imported from the Continent.

8

Irons (*left to right*): early 19th-century cap-iron, the screw hole simply punched through the metal: early moulding plane iron, probably 17th-century; fielding plane iron by William Crosbe (early-mid-18th-century); early double iron, the cap-iron located by a loose nut in the thickness of the iron proper; ordinary 18th/19th-century beading plane iron; unusual 19th-century iron marked 'Wm. Bingley Patent'. The patent apparently refers to the method of attaching the steel to the iron body (the join is clearly visible with a straight line). A Bingley chisel has also been seen, made of bronze with a steel face presumably attached by the same method. (*Old Woodworking Tools*)

RICHARD MELHUISH LIMITED, LONDON, E.C.

PLOUGH PLANES

No.		
78w.	Wood Stop, 6 Black Irons ...	**21/-**
79w.	Screw Stop, 8 ,, ...	**30/-**
80w.	Improved Side Stop, Double Hooped Stems	**42/-**
81w.	Ditto, Skate End	**45/-**
82w.	Ditto, Beech Screw Stems ...	**46/-**
83w.	Ditto, Box ,, ...	**54/-**

The last four numbers have eight best bright Irons.

Nos. 82w and 83w may be supplied with the following extras : Box Fence, **4/-** ; Handle, **12/-**

MOVING FILLISTER

No. 84w. With Brass Slip Stop. Price **11/-**
No. 85w. Do., with Tooth, and boxed. Price **13/-**
No. 86w. Brass Screw Stop, Forked Tooth, and Shoulder boxed. Price **20/-**
No. 87w. As above, but with improved Fence. Price **21/6**

No. 88w. With Wood Stop ... Price **13/-**
No. 89w. With Brass Screw Stop. Price **19/-**
No. 90w. With Brass Screw Stop, and Capped Stems ... Price **24/-**
No. 91w. Shoulder Boxed, left hand, Forked Tooth, improved Stop. Price **36/-**

SASH FILLISTER

"T" REBATE

No. 92w. Price **8/-**
,, 93w. Ditto, with Box Face ... **12/-**
,, 94w. Compass... **9/-**
,, 95w. Convex ... **9/-**
All above Planes are 1½ in. wide.

COACH SMOOTH
No. 96w.
1¾ in. cutter. Price ... **7/6**

STOP CHAMFER PLANE

No. 97w. With Fence as shown. Price **13/6**
No. 98w. Without Fence. Price **10/-**

"V" CHAMFER PLANE
No. 99w. Price **6/-**
,, 100w. Do., solid boxed. Price **7/-**

SHUTEING PLANE

No. 101w. Length, 22 in. ; 3 in. cutter. Price **27/-**

No. 102w. As above, Skew mouth. 3 in. cutter. Price **31/-**

SHUTEING PLANE

No. 103w. Length, 22 in. ; width of cutter, 3 in. With Plated Sole, Skew Mouth. Price **48/-**

MITRE PLANE
No. 104w. Length, 12 in. ; 2¼ in. Iron, Skew Mouth. Price **11/-**

JACK PLANE
No. 105w.
Single. 1⅝ in. Iron ; length of Plane, 10 in. Price **7/9**
No. 106w. If with Ram's Horn instead of Peg. Price **10/-**

MELHUISH'S PLOUGH
No. 107w.
New pattern Plough Plane, with quick-acting adjustment by wing nuts and wood screw stems. Brass Screw Stop. Price, with 6 cutters, **35/-** each.

ROUNDERS
(Wheeler's)
No. 108w.

Tapering :

⅞ × ⅝	1 × ¾	1¼ × 1	1½ × 1¼ in.
Price **10/6**	**10/6**	**11/3**	**12/-** each.

FORK SHAFT ROUNDERS
No. 109w.
Plain ... ... **21/-** each.
Brass-plated ... ... **31/6** ,,

Horticultural and Estate Appliances

The first of these is usually, like the smaller panel planes, not much more than about eight inches long, has a skew iron, a depth stop and either a fixed or a moving fence of the type used on side fillisters — a piece of wood attached to the sole by two screws in brass-lined slots. The iron and sole are often so shaped that the plane cuts not merely a rebate, but also the moulded part round the edge of typical eighteenth-century panels. This is most often a wide bevel separated by a shallow step from the main surface or 'reservation' of the panel, although sometimes a curved section was used. Panels of this kind, which went out of fashion around the end of the eighteenth century, are referred to as 'fielded', and panel-raising planes are often known as 'fielding' planes. Because of the change in fashion around 1800, most of these planes that are found are of eighteenth- or early nineteenth-century date, usually with the flat chamfers characteristic of the period. By the time of the Queen Anne revival in the last quarter of the nineteenth century, when fielded panels re-appeared in architectural joinery, machinery was taking over and fielding planes of this later period are not often found.

The badger plane was probably a nineteenth-century innovation, and is said to have been the invention of a plane-maker called Badger;

10

Chamfer planes: on the left is a home-made version of the Nurse type, in which a boxwood box (sometimes iron-soled) is adjustable to control the depth of chamfer. In the centre is a pattern often seen with the Preston name, in which adjustment is by a moving fence, and the bullnose position of the iron enables stopped chamfers to be made. On the right is a less common design, by John Moseley and Son, which dispenses with the traditional cutter wedge.

it is usually of jack size, and has a skew iron which is also tilted sideways so that the right-hand end of the cutting edge emerges at the extreme edge of the sole. The rebate of the panel plane is thus not required, and there is no restriction on the depth of the rebate in which the plane will work. Many badger planes have a slight shoulder along the right-hand side of the stock, like that found on moulding planes. Some also have boxing on the same edge of the sole, where the wear is greatest. Normally this is in the familiar boxwood, but other woods are found, including ebony.

Shooting, or chute-board, planes are used to trim mitres or square ends on timber held on a board which has a deep rebate in which the plane runs on its side. Such planes are usually of jack or trying size, and sometimes have a handle about halfway along the stock on what would be the left-hand side of a normal plane, so that it is uppermost when the plane is in use on the shooting board. Others have no handle at all. Many have a single iron (there is little point in having a double iron on a plane designed to work on end-grain) and with this often goes a distinctive form of wedge with a 'bulge' at the top, rather like that found on moulding planes. Shooting planes are found in relatively small numbers compared to the jack planes which they resemble in appearance, perhaps because an ordinary jack plane could always do the same job, provided that its sole was kept true and square to the right-hand side of the stock. Where much accurate shooting was required, as in a cabinet-making shop, more specialized planes were developed.

Most of these planes were of metal, and so do not fall within the confines of this chapter, but there was also a small form of shooting plane made of wood, and known as a mitre plane. It was very obviously related to the metal mitre plane, with a very low angle to the cutter such as is found in various metal planes but very seldom in wood planes. The stock is usually about twelve inches long, and the most distinctive feature is a boxwood block forming the front of the throat, which can be knocked down as necessary to take up wear on its lower end – which, being of end-grain boxwood, provides a very hard-wearing surface just at the most critical point. Whether these wood mitre planes, with their provision for preserving a fine mouth, developed as a cheaper alternative to the expensive metal variety or preceded the latter, I do not know; they appear, at present, to exist in smaller quantities. The type was certainly current in the early part of the nineteenth century, and indeed 10-inch and 12-inch mitre planes were still listed by tool dealers such as Nurse and Melhuish in the 1920s, although the absence of illustrations makes it impossible to see whether the wear-stop was still present, and also indicates that demand was limited. (Some late wood mitre planes had steel soles, and thus did not need the stop.) The larger, 22-inch shooting board planes (also sometimes referred to as mitre planes) continued to be listed (and sometimes illustrated) into the 1930s, both with conventional wedges

11

Chamfer plane: a very unusual form of adjustable chamfer plane in which the V-shaped sole is hinged to the stock. (A similar principle of adjustment was used on some very early Bailey iron planes.) (Christie's South Kensington)   (P)

12
Mitre, stair-rail, and strike-block. The wood mitre plane (top) has a typically fine mouth, with boxwood wear-stop. The plane on the right is thought to have been used in working stair-rails, and normally comes as a left- and right-handed pair. The strike-block plane at the bottom, by Buck of 242 Tottenham Court Road (post-1880), is a rare late survival of an early general-purpose plane used, for example, in trimming mitres before the advent of the mitre plane.
(Christie's South Kensington)   (P)

and the 'bulge' type mentioned earlier. (This was described in Nurse's catalogue as a 'scroll' wedge.)

Other derivatives of the larger bench planes include the cooper's jointer and sun plane and the gutter plane. The former is some six feet long, and would be quite impossible to use in the normal manner, so the mountain comes to Muhammad, as it were; the plane is mounted upside down with one end on the ground and the other supported on a simple trestle, and the workpiece (a barrel-stave) is pushed on its edge along the sole of the plane, to produce a true (and leakproof) joint. The sun plane, also known as a topping plane, is used for trimming the ends of the staves of a barrel after assembly, and is exactly like a jack plane except that it has no handle and is curved.

Gutter planes were used, as the name implies, for forming wood gutters; they look exactly like jack planes, but have a convex sole and cutting edge. Many that are found have clearly been adapted from ordinary jacks; this can be seen from the shape of the mouth, which will either be wider at the outer ends than in the middle, or will have had a piece of wood inserted and shaped to give a constant width. The same applies to any bench plane with a vertical front to the throat which has been converted to either a convex or concave sole.

Planes with concave soles are forkstaff or spar planes, the latter having the larger radius. These planes are normally of smoothing plane size, but with parallel sides: sometimes this applies only to the lower half, the upper being of the normal coffin shape. These tools were used in shaping handles, masts and spars and similar circular items. Again, converted bench planes are sometimes found, although the straight sides are an identifying feature.

Planes on which the sole curves from front to back are known as

13
German planes: the screw lever-cap smoother with mouth adjustment is often known as a 'Reform Plane' (Buck and Ryan, 1930).

compass planes. The wooden examples are nearly always convex only, and are used in trimming the inside of a curve – the inner edge of wheel felloes, for example. Their fixed radius rather limits their use, although some have an adjustable stop at the front, the lowering of which effectively increases the radius of the curve worked. However, this would necessitate lowering the iron as well, so that it would tend to tear at the grain, and the fact that the radius being worked increases with every stroke of the plane makes these planes less than easy to use. The stops may be of boxwood, sliding in a slot in the toe of the plane and sometimes tightened with a screw, or of iron, set into the forward part of the sole and adjusted by a brass thumbscrew on top. Wood compass planes with adjustable steel soles do exist, but not in large numbers.

One remaining smoothing plane derivative is the toothing plane; this has a near-vertical single iron, with serrations on its face which produce a toothed effect at the cutting edge. Often, they were supplied with more than one iron, with different serrations. Their purpose was to provide a 'keyed' surface on the groundwork of veneer, and also for initial trueing of uneven surfaces in knotty or awkwardly grained timber which called for a scraping rather than a slicing action.

Most of the remaining special purpose planes are of different form from those so far discussed, with narrow stocks and single irons; these comprise moulding planes, ploughs, fillisters and rebate (rabbet) planes. Because they receive less wear and tear than the bench planes, and because a joiner might need thirty or more different types, these tend to survive in larger numbers and from earlier dates. Most of the known English planes of the late seventeenth and early eighteenth centuries are moulding planes of one kind or another, and examples by the more prolific eighteenth-century makers such as Madox are not scarce enough to be beyond the reach of even a collector of modest means.

14
Forkstaff and Spar planes: the home-made example in the centre (in walnut) lacks the stopped chamfers of professionally made planes, and has a poorly formed mouth. The left-hand plane is also home-made (in fruitwood), but to a higher standard, and that on the right is in beech and may be professionally made, although it carries no maker's name.

15
Granfurdeus and Wooding: the Wooding ogee is
flanked by two very different planes with the oval
'Granfurdeus' stamp on the side, just visible on the
left-hand plane. That on the right, with its unusual
rounded shape to the upper part, is associated
particularly with Robert Bloxham of Banbury
(c. 1779).
(Christie's South Kensington)   (P)

Plane-making as a specialized trade appears to have been
established in the first two decades of the eighteenth century, and, as
mentioned earlier, was at first mainly confined to moulding planes.
Previously, joiners had made their own planes, and the best-known
plane-maker of this early period, Robert Wooding, was apprenticed as
a joiner, firstly (in 1693) to one Henry Shawe, and secondly (in 1699)
to Thomas Granford. The latter was also a joiner, but more
importantly, was by 1703 advertising his business as a maker of
planes and other tools. A number of early moulding planes exist with
a stamp on the side (not on the end-grain of the toe, as is usual) reading
'THIS IS GRANFURDEUS MAKE', and it is thought that this is a punning
reference to Thomas Granford's father, Henry the elder, made for the
benefit of his son (also Thomas), for whom Henry Granford would
have been 'Granfer'.

Robert Wooding, who died in 1728, was certainly in business as a
plane-maker, rather than a joiner, by 1725 and probably some years
earlier. Wooding was based in Queen Street, Cheapside, but another
early maker, Richard Elsmore, was making planes in Long Acre in
1713–15. With such a short recorded period of activity, it is not
surprising that Elsmore's planes are not often found, but he is
interesting as the first of a long line of plane-makers in the Covent
Garden area.

The moulding planes of this period are usually between ten and
eleven inches long, and are characterized by wide, flat chamfers along
the top and down the sides to the level of the shoulder or, on the flat
side, even below it. The shoulder varies in width according to the
width of the moulding, and does not exist on the narrowest hollows

and rounds, but it is normally concave, providing a comfortable rest for the fingers of the right hand, and on early planes usually slopes at a considerable angle. During the eighteenth century, the slope was gradually reduced until, more often than not, it disappeared altogether. Variations occur in this, as in the shape of the chamfers and of the decorative details at the end of the chamfers, between different makers' planes and even between planes of different periods by the same maker. Typical examples of this are well set out in Goodman's *British Planemakers from 1700*, which is the Bible for every plane collector. Occasionally, early planes are found with a different form of shoulder altogether, the decoration consisting of a small ovolo or ogee moulding which may or may not be returned at the ends. Not all planes were professionally made, and home-made examples are found from the nineteenth and even twentieth centuries, so that variations and anachronisms will always keep turning up.

From the time of Wooding and Elsmore onwards, it was common practice for makers to stamp their names on the end-grain at the toe of the plane, invariably on the upper, narrower part, until the advent of larger name-stamps (often including addresses and other information) in the latter part of the nineteenth century meant that this was not always practicable.

The 'Granfurdeus' stamp mentioned earlier is often well-nigh illegible, being stamped on the side-grain, but stamps on the end-grain are very clear and can still be read with ease after 250 years — unless, of course, they have been tampered with. There is, perhaps, no reason why anyone should want to tamper with a maker's name but many owners, either deliberatley or accidentally, struck their own names directly on top of the maker's, leaving it illegible. A further

16
Two 'hunchbacks' and two Bottles: the pair of planes on the left, a type seldom found in Britain, are thought to have been used in wardrobe mouldings. The right-hand planes both belonged to members of the Kentish Bottle family: that on the left, dated 1739, is home-made in fruitwood, that on the right is dated 1776 and is by Wheeler (1760–80). (Christie's South Kensington)　　　　(P)

problem with early planes is their length. Towards the end of the eighteenth century, moulding planes became standardized at a length of $9\frac{1}{2}$ inches, compared with $10–10\frac{1}{2}$ inches in the period of Wooding. In the nineteenth century, tool-chests were made to accommodate the then standard size of plane, and old planes that were too long to fit these chests were simply shortened by their owners. This could be done by planing one or both ends on a shooting board (planes so shortened are said to be 'shot') or larger amounts could be removed by sawing. If it was the toe that was removed in this way, the maker's name went too. Occasionally, a planed toe still has a 'shadow' of the name stamp preserved in the colour of the wood, which may become quite clear when linseed oil is applied.

In the case of stamps obscured by overstamping, it is sometimes possible to identify the name from the overall shape of the stamp, if the plane is by a well-known maker, or there are other planes with similar but clearer stamps in the same kit. The former method is not easy for a beginner, since it relies solely on experience of planes seen. It can sometimes be very frustrating if, when a stamp is recognizable as a familiar shape, the identity cannot be recalled; if the initial letter of the surname can be deciphered, a glance through the relevant section of Goodman's list (which is alphabetical) may be enough to jog the memory.

The style of makers' (and owners') stamps is a subject for study in its own right. As it is covered by a chapter in Goodman, there is no need to go into detail here, beyond pointing out the most obvious chronological variations. In the eighteenth century, most stamps had embossed (i.e. raised) lettering in a sunken panel. The edges of the panel were most often decorated, rather like a piecrust, and these are referred to in Goodman as 'zig-zag borders', or 'zb' for short. However, some, particularly early ones such as those of Wooding, have plain straight edges ('line borders', 'lb') and some a combination of the two, with a line outside a zig-zag, or, rarely, the other way about. Lettering was usually in Roman capitals, but some makers favoured lower case with capital initials, and this style normally has a border in which the top line is stepped to follow the height of the lettering, a useful feature in identifying overstamped names (like warning and obligatory road signs, shaped differently so as to be roughly identifiable even when obscured by snow). Cursive script is not unkown, but is more often found in the incuse stamps of the nineteenth century. These consist of lettering impressed into the wood, and usually have no border. Such stamps first appear about the beginning of the nineteenth century, although many makers continued to use the older style until at least the middle of the century. Although very small lettering was sometimes still used (the Glasgow maker Lourie is perhaps the most obvious example) there was a tendency for marks to expand, both by using larger and more widely spaced letters and by adding addresses or even, as in the case of Greenslade of Bristol, lists of Exhibition awards. It is seldom, if ever, possible to determine exactly when a maker's name-stamp changed in style (and, in any case, more than one stamp may have been in use concurrently) but at least these changes enable planes from some makers to be placed in some sort of chronological order. Gabriel, for example, a well-known London maker whom Goodman dates to the period 1770–1816, changed at some stage from 'zb' to incuse, and so did the very prolific Griffiths of Norwich – not surprisingly, as this firm was active from 1811 or before right up to 1948.

Many of the late, large stamps were placed on the lower part of the plane's toe end simply because they were too big to fit on the narrow upper part. With that proviso in mind, it is otherwise true to say that a name stamped anywhere other than on the upper toe section will usually be that of an owner rather than a maker. Owners' names are very frequently stamped more than once, sometimes with different initials where a plane has passed through more than one generation of a family. It is very rare indeed to find a maker's name twice, although it is not entirely unknown to find two different apparent maker's names, usually because one of them was also a dealer selling other

17
Two makers' stamps: it is unusual to find two makers' names on one plane. George Darb(e)y (1750–84) and John Briscoe (c. 1785) were both Birmingham makers. Both stamps are typical of the 18th century in style, although that of the Darbey stamp, with letters of equal height, survived well into the 19th century.
(Christie's South Kensington)   (P)

makers' products. The distinction between a maker and a dealer is not always crystal clear; many big dealers offered planes of their own brand, which might have been made in their own workshops or under contract, or simply bought in with no maker's name, while it was also quite acceptable for a dealer to advertise his name on a plane bearing a maker's name as well (just as many car dealers today decorate the rear windows of cars they sell with their own stickers). This practice is not particularly common, however, and is confined mainly to planes of the late nineteenth and early twentieth centuries. A late example is a rabbet plane bought from Buck and Ryan about 1964, which has their name stamped on the lower part of the toe (even though there is plenty of room on the upper section). The maker's stamp, EMIR LONDON, puts the clock back over two-and-a-half centuries by being on the side of the stock.

What constitutes an interesting maker's name? Clearly, if you are lucky enough to find a plane with a Wooding, Elsmore or Granford stamp, you have a very worthwhile addition to any collection, but almost any eighteenth-century plane is worth considering. Planes by any of Wooding's immediate successors, William Cogdell, Jennion, Phillipson or Robert Fitkin (all of whom were apprenticed to the Wooding business) are desirable. William Cogdell and Jennion were active in the 1730–50 period, and John Cogdell from 1750 until the mid-1760s or later. Fitkin also worked in this period, up to the late 1770s, and John Fitkin's business ran from 1789–1827. Phillipson was in Wood Street, Cheapside, from 1740–60, and the business continued at the same address until 1765. Moving west across the capital, we find William Madox making planes from the middle of the century until 1775, and one of his planes is likely to be the first of this period that will come the way of a collector today. There were several plane-makers at the time in the same area of Westminster, including John Rogers, John Hazey and John Mutter, all of Tufton Street. Mutter moved to Covent Garden, and then back to Westminster when he took over Madox's Peter Street premises in 1775, but he later moved again to Covent Garden. After 1811, the business passed to another maker already established in that area, John Moseley, and many planes of the period immediately following this are stamped MOSELEY LATE MUTTER. There were various changes of name in the long history of this firm, but for most of the nineteenth century it was known as Moseley and Son. Towards the end of the century, Covent Garden was deserted for New Oxford Street and then Kentish Town (1900–1914). Various other Moseleys were also in the business, by far the most prolific being John Moseley and Son of Bloomsbury (1862–94) and of High Holborn (1894–1910). During this latter period, the business was a subsidiary of William Marples and Sons of Sheffield, and it was transferred to Sheffield and incorporated in the Marples factory in 1910. That business eventually became part of Record Ridgway Tools Ltd., so that the present day Record iron planes can trace their

18
Plane by Cogdell (1730–52): the flat-topped, ovolo-moulded shoulder is unusual, although the flat chamfers are typical of the period. The moulding, a Grecian ogee, is a later modification. (Christie's South Kensington) (P)

ancestry, if somewhat circuitously, back to the Madox planes of the mid-eighteenth century.

The plane-making trade spread to provincial towns and even villages in the eighteenth century, and the products of these local firms make an obvious collecting theme for anyone living in their area. Thomas Holt of Lewes is an example; active in the 1760s and up to at least 1780, his planes would be an excellent subject for any collector living in Sussex. In Kent, Michael Saxby of Biddenden made planes from as early as 1730 for many years, and these are very collectable, if not exactly common. Kentish collectors can always 'fill in' with the plentiful Charles Nurse products, even if they confine themselves to those made in Maidstone, before the firm moved to Walworth Road in 1887, and there are also the planes from Crow of Canterbury made during the latter half of the nineteenth century. Collectors in Yorkshire are particularly well supplied, with Varvill, in various nomenclatures, covering the entire nineteenth century and other names like Bewley of Lees, Heathcott of Sheffield, John Green of York and King of Hull. The last, again with many changes in the name, ran from 1744 to 1907. Other notable plane-making centres were the Bath and Bristol area and the major Scottish cities.

The Scottish makers are particularly interesting, as many of their planes display characteristics not often found in English planes. They were keen, for example, on the use of two or even three separate irons, especially in wide cornice and other combination moulding planes. Mathieson of Glasgow (with addresses also in Edinburgh and Dundee) are particularly noted for their variety of ploughs and sash fillisters with 'extras' such as handles, screw-stems, bridle stems and prowed skates.

For some collectors, the chief interest in moulding planes lies not in who made them, but in the kind of moulding they make. The most common are perhaps not strictly moulding planes at all, but I include them under this heading as they are closely related in both form and function. These are planes for making grooves and rebates – ploughs, fillisters, rebate planes and housing planes.

The plough was the most elaborate tool in most old chests, boasting such refinements as an adjustable depth-stop, an adjustable fence and a choice of up to eight different cutters. The problem of making a sole that would suit eight different widths of cutter was solved by substituting a thin strip of iron or steel known as a skate (some even had the projecting 'prow' of the ice variety from which this component took its name). Although this supported the cutter only in the centre, the latter was exceptionally thick and rigid, and was positively held laterally by a groove engaging the leading edge of the rear part of the skate. This lack of a full-width sole would not make for a fine finish, but that seldom matters at the bottom of a groove.

The depth-stop normally consists of an iron-soled brass shoe projecting from the base of the stock adjacent to the skate, and

19
Twin-iron moulding plane: many of these are Scottish, although this one comes from just south of the border, being by Fitzakerley of Sunderland. It forms a bead and cove, or neck-moulding.

adjusted by a brass thumbscrew at the top, although early eighteenth-century models may occasionally be found with a simple wood depth-stop held in place by a wedge or by its own tight fit. The fence on British and American ploughs is normally attached to two stems which slide in holes in the stock and are held in place by wedges, or, on more expensive versions, by wood nuts and lock-nuts on threaded stems. Continental ploughs have stems fixed in the stock, with the fence sliding on the stems, which are often of the screw pattern. Although most British ploughs are of beechwood, some have certain parts in boxwood – all or part of the fence, the stems, stem wedges and, nearly always, stem nuts when present. Numerous other 'extras' can be found; the stems may have brass rings at the ends, held tight by wedges placed in the form of a cross in the end-grain, and some stems even have an inlaid boxwood or brass scale on top, graduated in inches. The depth-stop may have a clamping-screw in the side of the stock, working in a slotted brass plate. This plate normally has a decoratively shaped pointed top, giving such planes the sobriquet 'church window'. A closed handle may be provided at the heel, and sometimes the skate is extended at the front, the most attractive examples finishing in a scroll.

Adjusting the fence on these ploughs takes a certain amount of care, particularly as the fence needs to be kept parallel with the skate, and even the screw-stem variety is fiddlesome. Various attempts were made in the late nineteenth century to improve the system of adjustment, their one common feature being the use of metal at some point. Mathieson offered a plough with hollow brass stems, fixed in the stock, to which the sliding fence was clamped by thumbscrews passing through slots in the tops of the stems and gripping the inside of the hollow. The same firm is also associated particularly with the bridle plough. This likewise has fixed stems (usually of brass or ebony), to which the fence is clamped by a single thumbscrew passing through a bridging piece of iron or brass gripping both stems at once.

Perhaps the most ingenious system was that patented by D. Kimberley and Sons of Birmingham. Advertisements of the David Kimberley and Sons Tool Manufacturing Company Ltd. (short company names are a modern fashion) in 1902 claimed that 8,000 and more of these ploughs and fillisters were in use, but if that was so, they must have had a surprisingly low survival rate, for they are not often found today and are eagerly snapped up by collectors when they do appear. The Kimberley system uses metal stems, with a central screw-stem to provide a single-action adjustment, and all three metal stems are attached to an iron casting on top of the wood fence. Compared with the Mathieson types, the Kimberley ploughs have a rather mass-produced appearance, without the charm of some other mass-produced all-metal planes such as those of Preston and the earlier Stanley models.

Closely related to the plough is the sash fillister. This is a surprisingly common tool in the light of its specialized use, and no doubt indicates the important part played by window-making in most joinery shops into the early years of the present century. A fillister is a rebate plane with a fence, but the feature of the sash fillister is that it works a rebate on the side of the work-piece further from the fence: hence the need for a fence mounted on stems like those of a plough. The object of the design is to enable the rebate on a sash bar to be worked without having to turn the wood round after making the moulding.

A considerable amount of work was involved in the making of window sashes from scratch (particularly before the advent of plate glass and sheet glass in the mid-nineteenth century led to a great reduction in the use of glazing bars) and the saving of time gained by this method would have added up very advantageously over a period. An additional advantage was that the rebate would be true to the same edge of the timber as the moulding, so that the central, fat portion of the bar (which normally corresponded to the position of the mortice and tenon joints) would exactly match at all joints, and indeed there would be no need to true up the outside edge of the bars until the final trim after assembly.

Another approach to this problem was to use a 'stick and rebate' plane, which cut both rebate and moulding at once. The most sophisticated of these had a two-piece stock to give an adjustable gap between the two cuts, but although continental joiners favoured them, as did those in America and even Scotland, in England stick and rebate planes never found much favour.

Sash fillisters were available with the same range of extras and variations as the plough (as well as various forms of boxing for the sole) although most of them are less frequently found on fillisters. Two basic models were made, the earlier having its iron skewed back to the right so that, as on most fillisters, the forward end works into

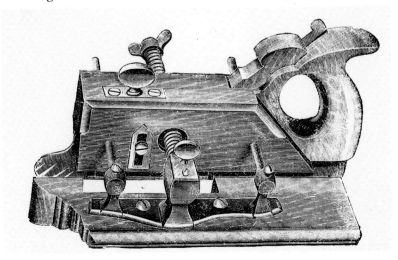

20
Kimberley patent plough: this illustration appeared in *The Illustrated Carpenter and Builder* in 1902, when it was claimed that over 8000 (including fillisters) were in use.

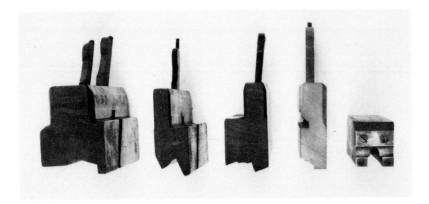

21
Sash planes (*left to right*): stick and rebate twin-iron plane by Arthington, Manchester (producing a bevel, or 'rustic', moulding); ovolo scribing plane; sash ovolo; late (1960s) sash ovolo, English, but of Continental type with equal thickness and no 'spring'; brass-tipped saddle templet for scribing ovolo sash-bars.

the angle of the rebate. The shavings in this case emerge to the right, falling on the bench. Although this type remained in production as a cheaper alternative, from early in the nineteenth century a new pattern was favoured with the iron skewed the other way, so that the shavings emerged to the left. Sash fillisters are sometimes referred to among collectors as 'on-the-bench' and 'off-the-bench'. Because the trailing edge of the skew iron of the latter works the edge of the rebate, a 'nicker' or 'spur' is provided to preserve a clean cut. This is a small cutter mounted vertically at the side of the stock in front of the main cutter, and scribes a line in the timber at the edge of the cut that is to be made. It is also found on planes which are likely to be used across the grain, and avoids the need for preliminary saw-cuts in housings.

For working rebates on the nearside of the timber, a side fillister can be used. This has a similar stock to the sash fillister, but without the

22
Dual-purpose fillister: Mathieson No. 18 combined sash and moving (side) fillister, with handle.

fence and stems; instead, a simple fence consisting of a piece of wood with two brass-lined slots is attached to the sole with screws. This plane is also called a moving fillister, by comparison with a fixed fillister, which has an immovable fence like that of many moulding planes. Fixed fillisters are not often seen, and when they do turn up they tend to be home-made. Where a fence is impracticable, or for enlarging existing rebates, an ordinary rebate plane is used; this comes in various widths from $\frac{1}{2}$-inch up to 2-inch, usually has a skew iron but seldom a boxed sole, and the throat is, as occasionally on fillisters, formed as a shaped opening right through the stock. There is no shoulder as on a moulding plane. For some reason, sash fillisters very seldom have the fence so designed that it can pass under the sole and thus enable the plane to be used as a side fillister. No doubt there is a good reason for the fence having its tip slightly above the level of the sole, as in most sash fillisters, but it is difficult to see what that reason may be, or why so few dual-purpose fillisters seem to exist.

Simple versions of ploughs also exist, most frequently in the form of a grooving plane for use in conjunction with a tonguing plane for working tongue-and-groove joints; these pairs are often known as matching planes, from their use in creating 'tongue-and-grooved matching'. They were made, like fillisters, with either fixed or moving fences, and the same applies to drawer-bottom planes, which are much the same as grooving planes but supplied singly, and designed to cut narrow grooves only – usually not more than $\frac{1}{4}$-inch wide. Some matching planes, especially on the Continent, combined both tonguing and grooving functions in one plane, with two cutting irons facing in opposite directions.

One further type of grooving plane remains, and that is the dado housing plane. This is used for cutting grooves or housings across the grain; its name derives from its original use in cutting a groove in the floorboards round the edge of a room, so that the panelling or dado on the walls could be housed into the floor and make a draught-proof joint. There is a double nicker iron near the front of the stock, for cutting through the grain on both sides of the groove, and an adjustable depth-stop. The wood sole is formed from the stock and extends the full width of the iron; there is no skate as in a plough or grooving plane. A seldom seen variation is specifically designed for the plane's original purpose. In order to keep the user's knuckles well clear of the wall, the stock is much thicker than the width of the sole dictates, and is fitted with a handle offset at an angle on the left-hand side. This type of plane is referred to in the 1899 Mathieson catalogue as a 'Flooring raglet', raglet being the Scottish term for dado plane.

Finally in this section, mention must be made of ploughs designed for circular work. These have a very short, curved skate and the stock is similarly cut away at the base but extends upwards and rearwards into a shaped 'tail' or handgrip. A beautiful refinement of this type of plough was that designed in 1846 by a Mr. Falconer, with an

23
Circular plough: this is a simplified version of the Falconer's Plough, with the fence sliding in a groove in the arm. Other models have a conventional wedge-stem.
(Christie's South Kensington)   (P)

adjustable flexible steel fence, enabling the plough to be used on 'circle on circle' work, where a curve exists in two geometric planes. The Falconer's plough, when it is found, tends to be made of ebony or a similar hardwood, and is a superb piece of workmanship; probably it was normally made during a man's apprenticeship, and seldom, if ever, produced by a tool manufacturer for stock.

Mouldings

Moulding planes form the largest single group of planes that are 'similar but different'. Until the 1930s, there was in architectural joinery a sort of grammar and syntax of mouldings and the way they were used. The most obvious example of this was the bead which always surrounded a door of any but the most humble variety, its purpose being to disguise the crack between the door and the jamb. The supply of ready-made lengths of moulding from the timber yard must have been partly responsible for the loss of this kind of practice. As it became increasingly unnecessary for a joiner to carry a large stock of moulding planes, and to work mouldings as a piece of work progressed, so it became more of an inconvenience to work or 'stick' even the simplest of mouldings in the solid. The so-called functionalist theory of design also helped, by giving a respectable excuse for omitting details that a previous generation would have considered essential if only to satisfy the tradesman's own pride in his work. In fact, many mouldings were themselves 'functional', as in the case of the door beadings mentioned above, but now that mouldings are beginning to come back into fashion, this aspect has been quite forgotten. (Just look at all those pseudo-colonial panelled front doors appearing on terrace houses and suburban semis all over the country, with raised or recessed panels carefully designed to catch all the moisture they can and store it there until the wood rots away. Before flush doors were invented, external doors not protected by a porch had for years had flush-fitting panels, at least on the lower half, with no more than a bead or reed around them, so that the water ran straight off.)

24
Falconer's circular plough: based on an 1846 design of a Mr Falconer, this high-quality plough is usually made of ebony or rosewood, and it has a flexible steel fence, adjustable by a screw which pulls the ends back like a bow-string.
(Christie's South Kensington)   (P)

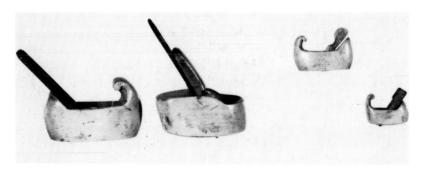

25
Violin-maker's planes: usually made of brass or bronze, these tiny planes are oval or circular in shape, often with slightly convex soles. These four are all made of bone: the largest is 2 in. long, the smallest $\frac{7}{8}$ in.
(Christie's South Kensington)   (P)

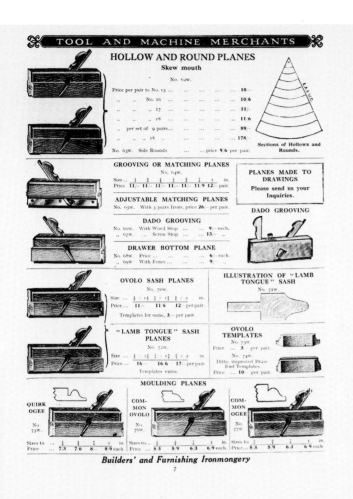

26

Hollow and round, grooving and moulding planes (from Melhuish's catalogue, 1925).

27

Planes for trimming mouldings: all the planes in this group are used for cleaning up or correcting mouldings. They were normally supplied in left- and right-handed pairs. Top to bottom: pair of side half-rounds; side rebate; snipe bill (its pair is at the bottom); side-snipe; side round. The side-rebate is probably 18th-century, with typical non-rounded chamfers and round-topped wedge.

Although one sometimes hears the proud owners of an old tool-chest claiming that they have a 'complete set of moulding planes', there is no such thing. The variety of moulding planes made over the years is infinite, and allowing for all the 'one-offs' made both in the workshop and professionally to order, no collection could ever be truly complete. This, of course, makes moulding planes an ideal subject for collecting – there is always something left to discover.

However, the basic element of a joiner's outfit of planes is a set with a definite limit: a set of hollows and rounds. A full set consists of eighteen pairs (each pair consisting of a hollow and a round), numbered from 1 to 18, the sizes ranging from about $\frac{1}{8}$-inch to $1\frac{1}{2}$-inch. Larger sizes were also available (usually 19 to 20) but were not included in the standard set. Many joiners bought a half-set, consisting of all the even numbers or all the odds. Full sets are quite difficult to come by, and even a full half-set may take some time to

find. Unlike most moulding planes, hollows and rounds have no fence and no depth-stop; they are used for general shaping or trimming of mouldings rather than for producing a specific profile, and are exceedingly useful in restoration work, where a new piece of moulding has to be matched to an existing one, and the chances of finding a plane that will make a perfect match are very remote. Hollows and rounds are so called after their own shape, unlike most moulding planes, which are known by the shape they produce; since they are different anyway, in not producing a specific moulding, this seems reasonable enough to me, and I do not follow Goodman's view that a hollow plane should be called a round and vice versa.

Companions to hollows and rounds are snipe-bills, side-snipes, side-rebates, side-rounds and side half-rounds. All these are used for 'free-hand' trimming and cleaning of mouldings, and come in pairs. Side-rebates and side-snipes have vertical irons, skewed across the

28
Two ploughs: on the right is a plough from the period 1730–50, by Cogdell, with a simple wooden depth stop and less ornament on the fence than later examples. On the left is a typical Mathieson product, incorporating their own design of hollow brass stems as well as a handle and projecting or 'prowed' skate. The contrast in colour between this late 19th-century plane and its 18th-century counterpart is due only in part to the difference in age; the more expensive Mathieson models like this were varnished originally and thus tend to retain their pristine appearance for longer.
(*Old Woodworking Tools*)

Two sash dowelling boxes: the joints between glazing bars were normally dowelled until about the mid-19th century, and these special jigs were used to ensure that the holes for the dowels were bored true. The left-hand example is by Sims, a Westminster plane-maker active in the 1820s, and that on the right, in mahogany, is home-made. In the Sims version, the sash bar is clamped in place by two parallel wood jaws in a tapered housing, which come together as they are drawn forward by the thumbscrew on the front. The position of the hole is adjusted by the screw on top. Sash boxes of any kind are quite rare.
(*Old Woodworking Tools*)

plane so as to provide a cutting edge at the side rather than at the bottom. The side-rebate is more robust-looking, with the stock tapered on the non-cutting side to a narrow sole, while the side-snipe has a boxed sole of ogee profile with a very narrow section at the lowest point, allowing it to be used in the narrow space above the quirk of a moulding. A snipe-bill is of similar design but with the cutter at a conventional angle, and emerging at the bottom to trim the quirk itself. Side-rounds normally form a quarter-circle each, one facing to the left and the other to the right, while side half-rounds have a flatter curve, slightly offset to the right and left.

Of the specific moulding planes, the most common are side beads: these normally came in sets of eight or nine, or were bought individually. The outside curve of the bead is continued downwards

30
Pair of radius and compass torus beads: probably used in staircase work, this unusual pair bears the stamp of Griffiths of Norwich, and was probably made to order.

on the stock to form the fence, and the iron is set into a housing in this part, to form a smooth transition from the bead to the flat edge of the work-piece. Better quality bead planes are boxed, usually for the full width of the bead and quirk on the smaller sizes and on the quirk or the quirk and the outer edge on larger ones. Unlike most other moulding planes of British origin, side beads are nearly always designed to be held vertically in use, as are centre beads. Other mouldings are made with the plane held at an angle; this is called the 'spring', and enables the shaving to clear the throat more effectively. Why it was not used on beads is difficult to see, particularly as it is sometimes found on torus beads, which are simply side beads set back from the edge of the timber and ending in a wide fillet at a lower level than the quirk. On many side beads, the depth-stop is made of a separate slip of beech held in a rebate in the stock by screws; such planes are described as 'slipped', and the purpose of this may be to allow for wear or, more probably, to enable the bead to be formed as part of a composite moulding where there would not be room for the depth-stop.

Centre beads have no fence, and normally have the bead in the centre of the plane's stock, with a depth-stop on both sides. There are surprisingly few applications for this moulding in use, and centre bead planes are accordingly much less common than side beads. Also fairly uncommon are cock-bead planes. These are normally of small size, for forming the narrow bead that was often fitted round drawer fronts in the eighteenth and nineteenth centuries. This bead projected from the surface of the drawer-front, and was applied, or 'planted' rather than 'stuck'. The plane has a central profile like the centre bead, but there are no quirks and no depth-stops, the sole being flat with a central flute to accommodate the bead.

Even rarer than the cock-bead plane is the cock-bead fillister. This is designed to cut the rebate into which the cock-bead fits on the bottom and sides of the drawer, working down from the face of the drawer-front. It therefore cuts a very narrow but relatively very deep rebate, and has a nicker to cope with the end-grain at the sides of the drawer. (The top cock-bead covered the full width of the drawer-front, and so required no rebate.)

An astragal is a form of bead; like the cock-bead, it stands proud of the surrounding surface, but it has a fillet on each side. It is used more in furniture than in architectural joinery, for covering the join between meeting doors, forming margins round panels, and for the glazing bars in glass-fronted bookcases and the like. (Glazing bars, even in windows, were often referred to as astragals in the eighteenth century.) Where astragals occur in architectural mouldings, they tend to be combined with other mouldings (the cove and astragal sash moulding, for example, in which the astragal is a very stylized version of its normal form). The relative scarcity of simple astragal planes may be due to this lack of architectural application, for in general moulding planes seem to have been used far more by the joiner than the cabinet-maker: this reflects both the greater quantity in which architectural mouldings would need to be made, and the fact that the

31
Mouldings (from Nicholson's *Practical Builder*, 1823): A, sash bars (cove-and-astragal, rustic, cove-and-astragal and cove-and-astragal stile); B, quarter-round or common ovolo; C, cove or cavetto (often wrongly called a scotia, cf. J); D, quirk (or Grecian) ovolo; E, quirk ovolo with bead; F, Grecian ogee (described by Nicholson as 'semi reversa'); G, cyma recta; H, ditto, Grecian; I, cyma reversa; J, scotia.

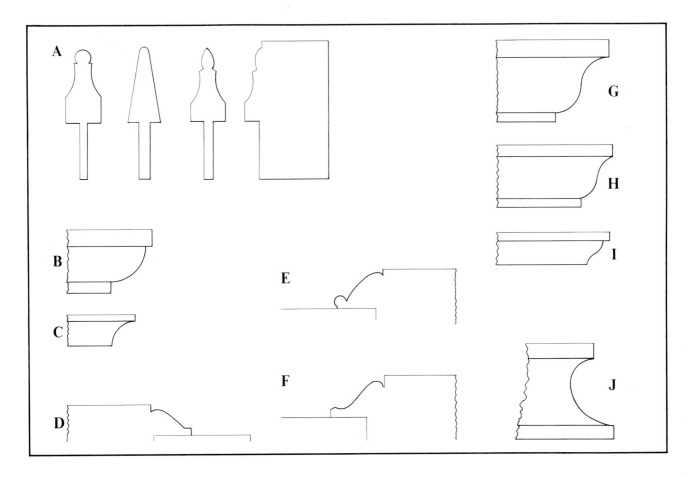

32
Reeding plane: only one reed is worked at a time, the left-hand profile being a removable 'fence'. The maker is Varvill and Son of York (1818–40).

cabinet-maker was often working in hardwoods which responded better to the scraping action of the scratchstock than the slicing action of the plane.

When two or more beads are grouped together, they are referred to as reeds. Reeds were particularly popular in the Regency period, Sir John Soane being especially fond of them, usually in triple form. Large reeds would, like most wide mouldings, be difficult to work, and would require less energy if worked as a series of centre beads. This would need the re-setting of a guide clamped to the work-piece, as there was no fence on the plane. An ingenious alternative was a plane with a double-reeded sole but a cutter in one section only, the 'dead' half acting as a fence by riding on the bead already formed by the previous cut. This section is removable so that the first bead can be worked with the same plane. I have never used this plane, but would imagine a fair amount of care must be needed to keep it in line during the eary stages of the second and subsequent cuts.

Some small reeds are arranged so that the centre is higher than the outer sections; these are known as cluster reeds, and, being essentially furniture mouldings (they are often found in picture frames), there is usually little or no perceptible quirk between each bead. It is unusual to find a quirk on any moulding in a piece of furniture, presumably because the main object of the quirk as found in softwood joinery was to allow for the filling effect of successive coats of paint. Stripped pine is, of course, a twentieth-century fashion.

One problem for the newcomer to this subject is the variety of names used for mouldings; many of these names are synonymous, but some are applied to different mouldings by different authorities. The simple quarter-circle concave profile may be called a cove, a cavetto, a covetta, a scotia or a quarter-hollow. Scotia is the name most in use by timber-merchants today, although strictly speaking a scotia is a sort of asymmetric flute; its name comes from the Greek for 'shadow', and in its original form, as found in the moulding at the base of a classical column, the slight return of the curve casts a shadow. Even without the return, a scotia ought perhaps to be elliptical rather than circular in section, but in practice it often is not. In general, mouldings with elliptical curves are 'Grecian', while those based on the simple circle are 'Roman'. The Grecian variety became very popular at the end of the eighteenth century, with the fashion at that time for all things Greek, following the publication in 1789 of the most important volume of Stuart and Revett's monumental work, *The Antiquities of Athens*.

Mouldings combining a convex and concave form are known as ogees, or by the latin term *cyma*. With the convex part at the top, the moulding is a common ogee or *cyma recta*; the other way round, it is a reverse ogee or *cyma reversa*. The 'top', in this context, is the face of the board on which the moulding is worked; if inverted (as it would be in a cornice) a *cyma recta* is still a *cyma recta*. From the common ogee is derived the quirk ogee, often found round door panels and architraves. The Grecian form of this moulding invariably has a slight

33
Grecian ogee with bead: a set of three graduated complex moulding planes by Moon (at the Little Queen Street address, 1832–51).

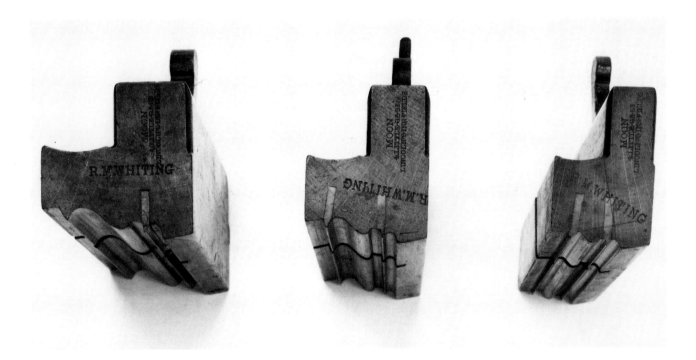

return at the base, usually stopping at a bevelled fillet. In England this moulding is normally called simply a Grecian ogee, without reference to the fillet, and was the most popular moulding for door panels and simple architraves from the late Regency period up to the First World War in Great Britain. I have never seen it used in eighteenth-century work, and although eighteenth-century planes are sometimes found with Grecian ogee profiles, close inspection usually shows these to have been adapted at a later date. The earliest evidence of the Grecian ogee that I have seen so far is in Nicholson's *The New Practical Builder and Workman's Companion*, published in 1823. The name given to it there, however, is 'semi reversa' – even though it is closer to a *cyma recta* than a *reversa*.

An ovolo, as the name implies, is usually egg-shaped or elliptical, although it is also often found with a more or less circular section. In the eighteenth century it was used around panels (often with only one fillet, especially in the earlier part of the century) but it was also used widely in window sashes, and continued to be so used up to the mid-twentieth century. Like other mouldings, it tended to become wider in the nineteenth century. Early eighteenth-century glazing bars were very thick, and the ovolo mouldings on them correspondingly large, but by the end of the century, it was quite normal for the thickest part of the bar to be no more than $\frac{5}{8}$-inch thick, and ovolo mouldings were therefore more elliptical than before. In the late Regency and Victorian periods, this was accentuated as the moulding increases in width towards the glass but could not increase in depth: already, the fillet between the two mouldings was barely more than $\frac{1}{8}$-inch wide. On many sashes of this period, the fillet was done away with altogether, the two ovolos coming to a point; this style of moulding was called Gothic, although the outer fillet was retained on the stiles and rails.

Although the most popular, the ovolo was not the only moulding used in sash work. The next in favour was the cove-and-astragal, an attractive late eighteenth-century profile which is often almost unrecognizable today after 200 years or more of regular painting have concealed its delicate lines. Like the Gothic style, this moulding has no quirk on the bars, which are therefore shallower than the stiles and rails. In any case, the mouldings on the bars were slightly different from those on the outer frame; because of the extreme thinness of the bars, the moulding was made slightly shallower, and sash planes were supplied in pairs, marked on the heel '1' and '2' respectively, to provide for this. After many years of use, the difference between the two planes in a pair is often difficult to discern

Where possible, the joints between the moulded parts in sash frames were scribed – that is, the end of the member with a tenon was shaped to fit over the moulding on the morticed part. To assist in this task, sash templates were used – beech blocks shaped to fit closely over the moulding, with the end cut to form a guide to the scribing gouge. Sometimes the ends were reinforced with brass, and some templates, known as 'saddle' templates, were double-sided to fit over a glazing bar and provide a more accurate guide. An alternative method of scribing was to use a scribing plane; these are not very often seen, but would have been useful where a large number of bars were to be scribed at once. The cut needed to be made before the mouldings were worked, when the bars were still rectangular in section and could be cramped tightly together so that a plane cutting into the end-grain would not tear out the fibres. Scribing could be used for ovolos and also for the so-called lamb's tongue moulding (actually a flat form of reverse ogee), but would not work on cove-and-astragals or any other mouldings finishing in a bead or a point. These had to be mitred.

The ability of the ovolo to be scribed no doubt accounts partly for its revival as a panel moulding in the machine-made doors of the 1920s onwards, until panels finally disappeared in the 1950s. Latterly, the earlier form, with one fillet only, made a reappearance as the distaste for any sort of ornament reached its zenith.

Between the ovolo of the eighteenth century and the Grecian ogee of the nineteenth, there was an intermediate stage when common ogees, usually with a quirk, were used, and equally popular were quirk and Grecian ovolos. Although some catalogues list these as two distinct varieties (in one, the Grecian ovolo has no fillet), in general the distinction is far from clear and it is probably better to think of the two as different names for the same thing. Many of these ovolos had an astragal or bead in place of the fillet, and even a double or triple reed is not unknown. This was a foretaste of the elaborate combination mouldings loved by the Victorians. These form a splendid basis for moulding-plane collecting, but many of them are very wide (some, especially those made in Scotland, have two or even three irons) and are almost impossible to use. Our ancestors may have had stronger arms than we do, but even they found these planes hard work; a hole is sometimes provided near the toe, for inserting a draw-rope to be pulled by an assistant. It is ironic that mouldings went out of fashion just as the means to make them without such vast expense of labour became available.

2 BRITISH METAL PLANES

Metal planes are known to have been in use in Roman times, but evidence of such planes in the ensuing centuries is limited. In the medieval period, small iron or iron-shod block planes were used by instrument makers, and several beautifully decorated small iron planes are known from the sixteenth and seventeenth centuries. Larger planes in metal also existed (an Italian example dated to the mid-sixteenth century is illustrated on page 12 of Roger K. Smith's *Patented Transitional and Metallic Planes in America*) and from these evolved the typical English mitre plane of the eighteenth century. As with wood planes, English makers gradually dispensed with foregrips and non-functional details such as the lowered sides in front of the throat.

Traditionally, these and other types of metal plane derived from them have been referred to as 'English Metal Planes', which is natural enough when one considers that Englishmen have long used this epithet of anything pertaining to the British Isles in general. Now that

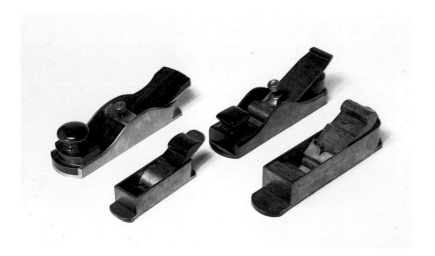

34
Mitre planes: *left to right*: Henley Optical Company plane, modern; small iron plane by Buck, 247 Tottenham Court Road (1867–79); Norris 'Improved Pattern' mitre plane; typical late 18th- early 19th-century iron mitre plane, with stamp of Iohn Green of York, with beech infill instead of the more usual rosewood. (The spelling 'Iohn' for John was normal at this time.)
(*Old Woodworking Tools*)

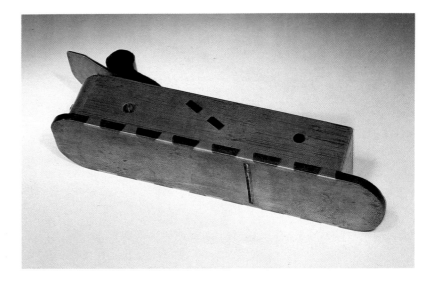

its use tends to be more specific, however, I have adopted the adjective 'British' for the title of this chapter, to take account of the large part played in the development of such planes by Scottish makers.

Initially, the technique appears to have been applied to mitre planes alone, for which a simple box-like form was apparently standard by the end of the eighteenth century. In nearly all cases, the wood infill was exposed at the top, and there was no form of hand-grip beyond the fact that the heel was rounded. The low-angle cutter was set at about twenty degrees, since it was intended to cut end-grain. It was invariably single, there being no point in fitting a cap-iron when the sharpening bevel was uppermost. It was also usually provided with an abutment at the top end, for adjusting with the hammer; cutters so shaped are often described as 'snecked'.

The metal most commonly used in early mitre planes was wrought iron. This term is not always understood today, when it is associated with ornamental gates and railings, most of which are now made not of wrought iron but of bent mild steel. Mild steel is a very different metal from wrought iron. It is more easily worked cold, but is also more rust-prone, which is why the Forth Bridge needs constant painting, in contrast to earlier nineteenth-century bridges made of wrought and cast iron. At red heat, however, wrought iron is the more malleable of the two forms. It has a grain structure not unlike that of wood and the fibrous appearance of a broken end is very distinctive. The sides of some early planes also show the structure of the metal as a series of dark, irregular lines, which will not be found on either mild or tool steel, nor on cast iron.

The plates making up the 'box' of these planes were dovetailed together, as on many later metal planes. Although it is natural to

36
Wrought iron: this mitre plane body by Robert
Towell shows well the laminar structure of wrought
iron.

assume that this use of a woodworker's joint stemmed from the plane's natural ambience, dovetailing was not unknown in metalworking. It was used, for example, in copper cooking utensils, where sides and base were dovetailed and brazed together. In metal plane-making, the dovetails are so formed that they cannot be pulled apart in either direction. In a wood dovetail joint, the two pieces can be assembled in one direction only and, glue permitting, can be pulled apart again in the same direction. In metal, by forming the 'tails' with bevels in two directions at once, the pins can be hammered into shape after assembly, as in riveting, so that the joint is locked against movement in either direction. The forming of these joints was carried out to a very high degree of precision, so that they are often invisible; only where rust has formed in the joints, or where two contrasting metals are used, is the construction immediately clear.

The mouth in most mitre planes is very fine indeed, and to achieve this, the sole on early examples is usually formed of two pieces joined at each side of the mouth with a tongue-and-groove joint, this being easier than forming an accurate narrow slit in solid iron. The infill and the wedge retaining the cutter were usually of rosewood or mahogany (a combination of the two, with mahogany for the less visible rear infill, is not unknown and may not always be due to a later repair). More often than not, a moulding was formed on the rear end of the front infill forming the front of the throat. The wedge was retained by a fixed bridge or bar of flat metal across the stock, and this also was usually ornamented on its lower edge. Makers' names are often on this bar, particularly in the case of planes by Robert Towell, who specialized in metal planes; makers of wood planes sometimes favoured the traditional position on the toe (e.g. Ponder), but often preferred to use the end-grain wood for which their stamps were designed by stamping their name in the throat.

Little appears to be known of Robert Towell, who may be the first specialist metal plane-maker, or at least the first to put his name on his products. Mitre planes were certainly being made in the eighteenth century, but, judging from examples of his planes seen so far, Towell was clearly active in the first two or three decades of the nineteenth. Many of the planes bearing other names were probably made by him,

37
Frogatt mitre plane: detail showing the stamp of
this wooden plane-maker placed on the wood in the
throat of a cast-iron plane (B. Frogatt: 1765–90).
(Christie's South Kensington)   (P)

38
Metal planes (left to right, top to bottom): iron
shoulder plane of unusual shape; Norris A5
dovetailed steel smoother with Patent Adjustment;
Norris A1 dovetailed steel panel plane (20½ in. long)
with adjustment; rare Spiers skew-iron mitre plane:
standard bullnose rebate plane: iron shoulder
plane. Both the shoulder planes are user-made, and
have mahogany infills.
(Christie's South Kensington)   (P)

for it seems unlikely that many of the wooden plane-makers would
have had the facilities for so specialized and distinctive a craft.

Not all mitre planes used this dovetailed wrought iron construction; even early types are found with the upper section made of
bronze or brass, dovetailed to an iron sole, and a cheaper method was
to form the entire box as a casting in iron, or, in some late examples,
bronze. Cast iron does not have the same laminar structure as wrought
iron, but it is likewise less rust-prone than steel, and has a similar grey
colour to that of wrought iron. It is very brittle, and cast-iron planes of
any type are often found with cracks or pieces chipped off. Sometimes
the arris at the mouth is chipped, so that the fine mouth which is one
of the principal features of a metal plane is lost.

An undamaged cast-iron mitre plane is not always easy to
distinguish from the wrought variety, since neither the laminations

nor the dovetails are always clear in the latter. One useful indication on some cast-iron planes is a cove moulding at the junction of the heel or toe with the projecting sole. Even where there is no moulding as such, there will tend to be a slight rounding of the angle compared to the sharp ninety-degree angle found with a fabricated construction.

The great majority of mitre planes of this box pattern appear to date from the first half of the nineteenth century. They were still available, albeit to special order only, in the early years of the present century, and could be had either with the traditional (but by then outmoded) wedge or the pivoted lever normally found on Spiers and Norris bench planes of the period. These late examples are seldom found, and the same can be said of the 'Improved' pattern offered by both Spiers and Norris in their twentieth-century catalogues. These have several features of other later metal planes – the lever cap, the 'humped' sides and the square fore-grip of the panel and jointer planes. The low angle of the cutter does not admit of a conventional rear handle, but the

39
Shoulder and chariot planes: *left, top to bottom*: gunmetal shoulder plane by Arthur Price, iron shoulder plane by Spiers and an unusually decorative gunmetal example by an unknown maker; *right, top to bottom*: London pattern gunmetal bullnose shoulder plane by Holland, gunmetal chariot plane and small, steel-soled shoulder plane, both anonymous.
(*Old Woodworking Tools*)

40
Panel and smoothing planes: top to bottom: Spiers panel plane, 17½ in. long (the largest size of panel plane); 12½ in. panel plane with traditional wedge held in slots in the casting (no maker's name); smoothing plane by Slater, Clerkenwell, with wedge retained by pivoted bridge; traditional Scottish cast-iron smoother with screw-lever cap, marked D. Galloway, Edinburgh.
(*Old Woodworking Tools*)

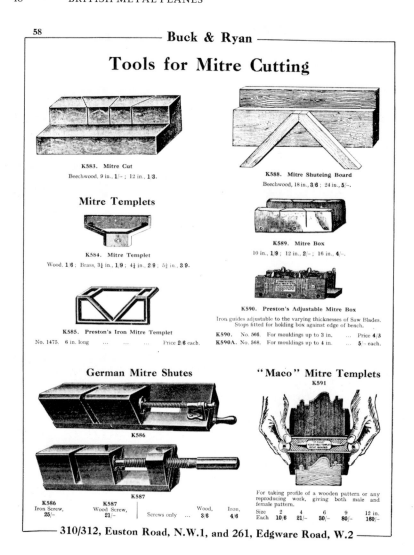

58

Buck & Ryan

Tools for Mitre Cutting

K583. Mitre Cut
Beechwood, 9 in., **1/-** ; 12 in., **1/3**.

Mitre Templets

K584. Mitre Templet
Wood, **1/6** ; Brass, 3¼ in., **1/9** ; 4¼ in., **2/9** ; 5¾ in., **3/9**.

K585. Preston's Iron Mitre Templet
No. 1475. 6 in. long ... ... Price **2/6** each.

K588. Mitre Shuteing Board
Beechwood, 18 in., **3/6** ; 24 in., **5/-**.

K589. Mitre Box
10 in., **1/9** ; 12 in., **2/-** ; 16 in., **4/-**.

K590. Preston's Adjustable Mitre Box
Iron guides adjustable to the varying thicknesses of Saw Blades.
Stops fitted for holding box against edge of bench.

K590.	No. 566.	For mouldings up to 3 in.	...	Price **4/3**
K590A.	No. 568.	For mouldings up to 4 in.	...	**5/-** each.

German Mitre Shutes

K586

K587

K586	K587		Wood,	Iron,
Iron Screw,	Wood Screw,			
25/-	**21/-**	Screws only ... **3/6**	**4/6**	

"Maco" Mitre Templets
K591

For taking profile of a wooden pattern or any reproducing work, giving both male and female pattern.

Size	2	4	6	9	12 in.
Each	**10/6**	**21/-**	**30/-**	**80/-**	**160/-**

310/312, Euston Road, N.W.1, and 261, Edgware Road, W.2

41
Various mitre-cutting aids (though the Maco was simply an adjustable templet, not specifically connected with mitres).
(Buck and Ryan, 1930)

infill is raised and shaped into a pear-shaped hand-grip. Skew irons were available at extra cost, and another interesting 'extra' listed in Spiers' 1909 catalogue was a traditional wedge and iron cross-bar – the extra cost perhaps reflecting the care required with this form of construction to ensure that the angle of the bridge and the wedge were exactly matched. Rarest of all the late mitre planes must be the Norris model of the 1930s which incorporated their patent adjustment for the cutter; curiously, this plane reverted to the old box form.

It is curious that a plane which, from the numbers surviving, must have been a fairly regular part of a cabinet-maker's kit in the early nineteenth century, became apparently less indispensable in the later part of the century. Possibly its specialized application was felt not to justify its cost when much of its work could be performed quite adequately by the other metal planes then available, which were of more general use. There would also have been a steady supply of secondhand mitre planes at that time.

Metal mitre planes were not necessarily used on a shooting board, as were most of their various wooden counterparts; rather, they were used in conjunction with a mitre vice, a sort of cramp with jaws cut at forty-five degrees to the base, so that a piece of timber held between the jaws and resting on the base could have its mitred end trimmed exactly true to the faces of the jaws. Like many tools made by cabinet-makers for their own use, these vices are often made of mahogany or other hardwood offcuts, although they could also be bought ready made, normally in beech with a wood or iron screw. The 'Improved' pattern mitre planes were clearly intended for use in this way, as their hand-grips would have been useless on a shooting board.

A reader eager for historical dates in the development of these planes will notice a sad lack of information. Between the mitre-plane period and the heyday of the Spiers and Norris planes in the late nineteenth century, there is a grey area when many metal planes were undoubtedly in use, but few survive with makers' names that enable them to be dated and thus to be placed in any sort of chronological order. Possibly the rebate plane was one of the earliest to appear after the mitre plane; examples by Towell are known, and it is the closest in style of all the metal planes to its wooden counterpart. A number of particularly attractive panel planes, which normally range in length from about twelve inches to eighteen inches, have been found in Scotland, suggesting that the larger British metal planes may have evolved there in the 1830s. This ties in with the story of Stewart Spiers, who as an apprentice cabinet-maker in Edinburgh bought a casting for a plane for 1s 6d, completed it and sold it (in his native Ayr) for 18s. This was about 1840, and led to the establishment of the Spiers of Ayr plane-making business.

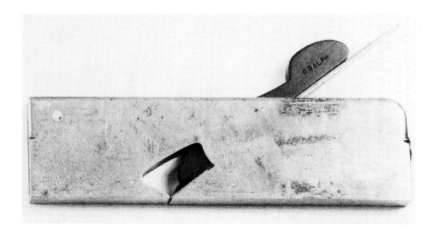

42
Rebate plane by Robert Towell: made of dovetailed wrought iron in the early 19th century, but differing very little from later steel rebate planes.
(Christie's South Kensington)   (P)

43
Scottish panel plane: made in gunmetal with mahogany handles and overstuffing. The scrolling fore-grip is reminiscent of earlier wooden Continental planes.
(Christie's South Kensington)   (P)

Most of the pre-Spiers Scottish planes (which no doubt continued to be made after Spiers began manufacturing) have cast stocks, usually in iron but sometimes in brass or bronze, and characteristic features include curved heels, projecting soles with cove mouldings and beautifully shaped 'stuffing', as the wood infill is often called. Most are 'overstuffed' – that is, the wood overlaps the upper edges of the stock so as to be flush with the sides, and there is sometimes an elegantly shaped scroll forming the foregrip. As these planes were probably mostly finished by tradesmen or, more probably, apprentices, there is little standardization of style or of wood. Walnut is often used.

Spiers planes which can be dated firmly to the 1840s and 1850s occur less frequently than one could wish, but the similarity between Norris and Spiers planes suggests that by 1860, when the Norris business is thought to have been founded, the main types familiar in both companies' output were well established. One feature that seems

44
Scottish panel plane: a late 19th-century example retaining traditional features such as the coved heel and toe and beautifully shaped walnut handle.

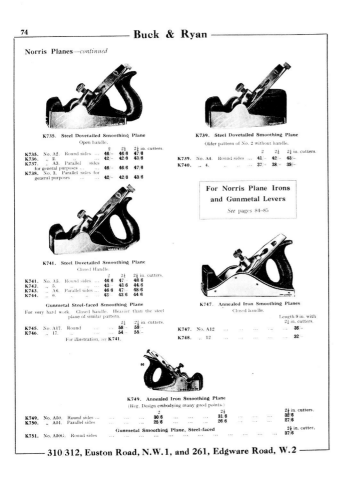

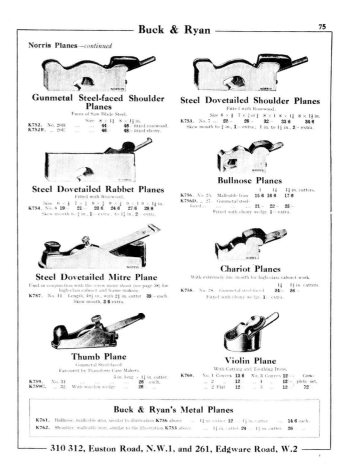

45
Norris planes, from the 1930 Buck and Ryan catalogue.

to be invariable is the use of rosewood for handles and stuffing, with the exception of ebony as a more expensive option on smaller planes such as rebate, shoulder and some smoothing planes. Ebony-filled smoothing planes are very few and far between, however, and may have been made only in unhandled models. Only in the twentieth century did beech or other native hardwoods come into use on some models; Norris usually disguised such substitutes with a coating of ebony or rosewood-coloured varnish stain. This does not wear well, and often cracks and flakes even on planes which have seen little or no use.

Bench planes were available in smoothing, panel and jointer sizes; the latter two were the equivalent of wooden jack and try planes, respectively. Smoothers came in a variety of shapes, broadly classifiable into those with parallel sides, partly tapering sides or curved (coffin-shaped) sides, and those with or without handles. Panel and jointer planes from Spiers, Norris and other professional makers showed less variety: most had a closed handle behind the cutter, a shaped square or rectangular 'knob' or foregrip at the front and, on larger sizes, a projecting sole at the heel and toe. As on the transitional Scottish planes mentioned earlier, all these bench planes had sides that rose to a hump (usually a double hump) alongside the throat, and tapered away in a curved line to front and rear.

Until the middle of the nineteenth century, if not later, the cutter was held in place by the traditional wedge, tightened either against a cross bar or in tapered slots cast in the sides of the throat, an inheritance from English wooden plane construction. However, the majority of metal bench planes have a screw lever-cap, usually of gunmetal (a kind of bronze). This cap is pivoted to the sides of the throat and cannot be removed from the plane; at the top is a knurled screw which bears against the top of the cap-iron. Tightening of this screw causes the wide lower end of the lever-cap to press hard on the cap-iron just above the cutting edge, where pressure is most needed. Unlike a wedge, the lever-cap is self-adjusting to the thickness and angle of the cutter and is not affected by wear in itself or other parts of the plane.

Used with a rigid cross-bar or side-slots, a wedge has to be a perfect fit if it is to remain firm in use. Two ways existed of overcoming this problem. One was to use a round cross-bar, found on continental wood planes, but seldom, if ever, on the British metal variety, which provided a less firm grip than was possible with the other methods. More interesting was the use of a pivoted cross-bar, which had a large enough flat surface to grip the wedge, and also automatically adopted the angle of the wedge as the latter was hammered home. Possibly this method gave rise to the lever-cap, which is an extension of the same idea; however, it is most often seen, in my experience, on planes by Slater of Clerkenwell, a metal-plane maker dated by Goodman to the period 1868–77. Slater seems thereby to have kept the wedge alive after other makers had largely discarded it in favour of the lever on metal bench planes. He also made planes with lever caps.

I mentioned earlier the variety of designs in smoothing planes, and it is interesting to see that in the 1914 Norris catalogue, in this category alone, there were some eighty permutations to choose from, taking into account all the sizes, shapes and materials. Prices ranged from 15s 6d for a basic, unhandled cast-iron model (with a choice of two shapes) to 28s for a de luxe version in gunmetal with steel sole, rosewood handle and Patent Adjustment. Even the cheapest of these would have represented a considerable outlay for a joiner or cabinet-maker in 1914, and it is perhaps surprising that so many of these planes exist.

The Patent Adjustment was at that time a new feature (the patent dated from the previous year) and consisted of a long screw behind the cutter with a knurled head at the top. This screw was mounted on a pivot so that it not only controlled the depth of the cut but, by sideways movement, the lateral position of the cutter as well. At the

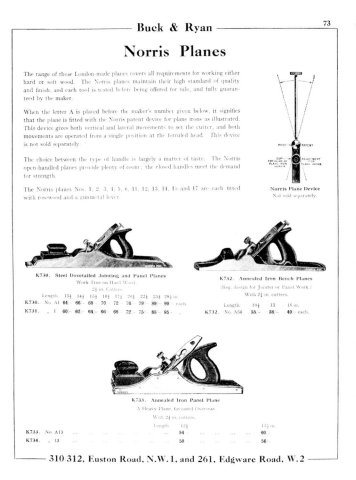

46
Norris panel and jointer planes, with details of the Patent Adjustment (Buck and Ryan, 1930).

47
Preston adjustable planes: of these two attempts to compete with the Bailey design, the upper is probably the earlier, and 'borrows' its handle from one of the same maker's conventional British metal smoothers. Both planes are rare.
(*Richard Maude collection.*)
(Christie's South Kensington photo)

lower end of the lever was a circular 'cup' which engaged the head of the cap-iron retaining screw. Although the lateral and vertical adjustments of the Bailey planes were thus combined in one lever, it was less convenient to use than the latter, since the depth of cut (the adjustment most needed once the cutter has been set initially) cannot be altered without removing the hand from its postion on the handle, and the screw on the lever-cap might need slackening if it had been tightened too much for the adjustment to work. (The latter problem does not arise on Bailey planes since the lever is held by a cam which always applies the same pressure.) Similar forms of vertical adjustment were adopted by one or two other makers (including Preston and Guaranteed Tools Limited, whose GTL bronze smoothers, otherwise similar in appearance to the Bailey type, are found quite often). However, the full Norris adjustment with lateral movement was not used by other makers, probably because, by the time the patent expired in the 1920s it was clear that the future lay with the Bailey pattern. Indeed Norris alone continued to flourish in the

48
Two user-made iron planes: the cast-iron smoother on the left probably dates from the mid-19th-century, and retains the traditional British method of retaining the wedge. It is overstuffed in walnut, with fine moulding on the fore-grip. The shoulder plane is of later date and, in spite of its rough appearance, is an excellent working tool with a sweet cutting action.

manufacture of these expensive planes until the outbreak of the Second World War.

The basic smoother in both the Spiers and Norris ranges was a derivative of the more box-like planes derived in turn from the old mitre pattern; it differed in that, in dovetailed versions, the wood infill was exposed at the ends and, in all models, projected above the metal sides. At the back it was high enough to provide a comfortable grip for the right hand, and the front, although lower, also formed a hand-grip. The sides were either curved (giving the plane the shape of a coffin; 'coffin smoothers' are of this shape, not planes for smoothing coffins) or were parallel at the front but tapered towards each other at the rear, again to make the plane more comfortable to hold. This shape was available from Norris in 1914 in cast iron, malleable iron or gunmetal; the coffin type could be had in all these materials as well as dovetailed steel, and with or without a handle.

The ease with which cast iron can be broken initially gave the expensive dovetailed steel planes a considerable advantage. Wrought iron planes by Spiers are seldom seen, and they may never have been made by Norris. The use of annealed, 'malleable' cast iron overcame this shortcoming more cheaply, but dovetailed steel planes nonetheless continued to be the most favoured. In 1914, a 2-inch Norris 14 (tapered heel type) cost 15s 6d in cast iron, 16s 6d in malleable iron and 21s 6d in gunmetal; the nearest equivalent in steel, No. 4, cost 19s. The gunmetal planes look very impressive, and were said to be particularly suitable for difficult work because of their greater weight. They normally had a thin steel sole. Whether they had any real practical advantage is open to question.

Malleable iron was initially offered as an alternative choice in existing cast-iron styles, but it seems that even before the First World War there was a need for a cheaper metal plane to meet the challenge of the American Bailey planes. Thus there appeared in the 1914 Norris catalogue a range of cast and malleable iron planes with closed handles and Patent Adjustment at prices similar to or less than those of existing models of similar sizes but without adjustments, or, in the case of smoothers, handles. The savings in cost were effected by using beech and walnut instead of rosewood, and that in smaller quantities. There was no infill at all behind the cutter; the front edge of the handle provided support for the cutter, and it was bedded in a raised formation in the base casting. In front, a short infill had a rounded top to provide a foregrip, and the fact that it was just that and no more was emphasized in the panel versions of the plane, where the same size grip was used and the extended casting in front of it remained innocent of wood infill. These planes (Models 50–53 initially; the panel sizes, 52 and 53, were later brought together in the catalogues as 54) were also made in gunmetal, but in none of their forms did they succeed either in ousting the American planes (which cost roughly half as much and had rosewood handles into the bargain) or in displacing the more traditional and handsome Norris tools. The exposed, unfettled inner parts of the casting and the stained handles have a rather cheap appearance (a few examples have been observed

49
Norris 50G smoother: made in gunmetal with steel sole. Although fitted with the Patent Adjustment, the model number is not prefixed by 'A' as there was no non-adjustable version.
(Christie's South Kensington)　　　　　(P)

with rosewood handles) but they are good working planes and are preferred by some users to the more traditional Norris models.

Spiers never produced an adjustable plane in significant quantities; even their 'Empire' plane, little more than a copy of the Bailey type, is rarely found (probably because it came at the end of the firm's effective life, about 1930). However, they did produce a cheap, non-adjustable plane which is found from time to time, and carries on its cast-iron lever-cap the name 'Plane-o Ayr'. Like the Norris '50' range, there is no rear infill, and on some there is none in front, but a Bailey-type turned knob instead. Some examples are a simple casting, but others have a base casting with steel sides and sole riveted to it. Some later Norris planes also had cast-iron levers.

Special purpose planes in this category are for the most part not derived from wooden counterparts. Only rebate planes are obviously so related. These retain the basic shape of the wooden variety, even down to vestiges of the stop chamfering, and many even retained the traditional shape of the wedge. Only the cutter angle was different; at around twenty-five degrees, it had the bevel uppermost. Skew irons (almost standard on wooden rebate planes) were an optional extra on the metal variety, and one that was not often chosen. Widths ranged from $\frac{1}{2}$-inch to $1\frac{1}{2}$ inches, the majority being in the $\frac{5}{8}$- to 1-inch bracket. Construction is nearly always dovetailed, and commercially made examples have rosewood or ebony infills and wedges. A seldom found variation is the twin-iron rebate plane; this has an extra mouth just behind the toe, so that the plane can be used either as an ordinary rebate plane or as a 'bullnose' type, for working into corners. As shown in the catalogues, this type was supplied with an iron and

50
Spiers 'Plane-o-Ayr': a late Scottish metal plane which offers neither the facilities of the Bailey pattern nor the quality of the more expensive Spiers models.

wedge in each position, which seems unnecessary as only one would need to be used at any one time; certainly, surviving examples normally have only one iron and wedge, which can be fitted in either position, the second pair presumably having been put out of the way in an oddments box many years ago.

Much more variety is seen in the shoulder plane. This is intended for cleaning the shoulder, or meeting edge, alongside a tenon or housing joint. Being designed specifically for end-grain work, it has an even lower cutter angle than a rebate plane. The wedge is extended outwards behind the heel of the plane to provide a hand-grip (the body of the plane is difficult to grip when in use against a long tenon) and the top of the stock is invariably shaped to some sort of curve, usually with a raised and sometimes projecting 'prow' at the toe, to balance the projecting wedge visually.

As the illustrations show, there was no rule as to the extent of this shaping; the shoulder plane is perhaps the only British plane made in significant quantities which never achieved a standard decorative convention. Even the different kinds of metal smoother are related to the choice of materials, sole shape or other working details. Shoulder planes are found in dovetailed steel, cast or malleable iron and gunmetal, with or without steel soles in the latter case. As with rebate planes, infill is normally rosewood or ebony, but, again, homemade examples can be found with almost any wood used. Like other home-made metal planes, they may have been made from purpose-made castings, from the user's own patterns, from bought-in ready-made castings or from scratch using iron or steel plates riveted together. The latter are very seldom dovetailed, and indeed the method of fixing the sole to the sides is seldom clear.

Like the earlier mitre planes, shoulder planes exist in surprisingly large quantities for their rather specialized application. Some, particularly the more elaborately shaped, were probably apprentice pieces, but enough exist with known makers' names on the toe to indicate that, up to the 1930s, no self-respecting joiner or cabinet-maker considered his outfit complete without one – even though many appear to have had very little use.

RICHARD MELHUISH LIMITED, LONDON, E.C.

IMPROVED IRON PLANES, English Manufacture

BULLNOSE REBATE

No. 192w.

Nickel-plated, with Rosewood Wedge, 3 × ⅜ in.
Price **6/6** each.

BULLNOSE REBATE

No. 193w.

Japanned. Length, 3¾ in.; Cutting Iron, 1⅛ in.
Price **7/6** each.

PATENT ADJUSTABLE BULLNOSE REBATE

No. 194w.

Nickel-plated.   Quickly adjusted by means of knurled nut.
Length, 4 in.; Cutting Iron, 1⅛ in. Price **12/6** each.

SHOULDER REBATE (Registered)

No. 195w.

Nickel-plated, Rosewood Wedge.
Length, 5 in.; Cutting Iron, ⅝ in.
Price **12/6** each.

JAPANNED BULLNOSE

No. 196w.

Length, 3¾ in.; Cutting Iron, 1⅛ in.
Price **8/6** each.

SHOULDER REBATE

No. 197w.

Adjustable. Nickel-plated.

Sizes ...	...	5 × ⅝	8 × ¾	8 × 1	8¼ × 1¼	in.
Price...	...	**14/6**	**26/-**	**29/-**	**30/-**	each.

COMBINATION BULL-NOSE AND FILLISTER PLANE

No. 198w.

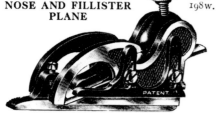

With Fence and Stop. Fence is attached to bed of Plane and is adjustable for varying depths of rebate. Both may be detached. Japanned.
Length, 3¾ in.: Cutting Iron, 1¼ in. Price **11/6** each.

ADJUSTABLE SIDE REBATE

No. 100w.

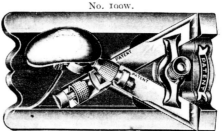

With right- and left-hand Cutting Irons. Nickel-plated.   Price **17/6** each.

Engineers' Tools and Machinery

53
Preston side-rebate plane: this differs from the smaller Preston model (and those of other makers) in having screw adjustment for the cutters. The round knob is relatively scarce, most examples having a more elongated form (cf. p. 58). (Christie's South Kensington)   (P)

52
(*Left*) Preston shoulder and rebate planes (Melhuish, 1925)

Most shoulder planes are about eight inches long, but the bullnose variety is only half that size. Of the two most commonly found patterns, associated with Spiers in one case and with Norris and other London makers in the other (it is sometimes described as 'London pattern') the latter is perhaps more elegant. The fact that its shape was so standardized among several makers suggests that it may be of earlier origin than the late nineteenth-century date of most surviving examples suggests. Some attractive all-metal variations were made by Preston, however.

This firm was known particularly for rules and levels, but also made planes of both the wood and metal varieties, and in the late nineteenth and early twentieth centuries they introduced a range of highly individual metal spokeshaves and special purpose planes which neatly filled the gap between the expensive British metal planes and the mass-produced products that were starting to flood the market from across the Atlantic. Bullnose planes of both the Spiers and Norris patterns were made, but more distinctive was their 'Preston pattern' bullnose rabbet plane (Preston, like Stanley, seem not to have recognized the conventional but slight distinction between a shoulder and a rebate plane). A version of this was even offered with fence and depth-stop, making it a bullnose moving fillister – a most unusual tool and one that is seldom found with fence and depth-stop still intact. This had an iron screw-lever to retain the cutter, but there was also a bullnose rebate and a 'shoulder rebate' plane of cast iron with ornamental (and weight-saving) recesses in the sides, which used the traditional rosewood wedge. Other Preston planes will be covered in the chapter on American planes, to which they are more closely related.

54
Shoulder and rebate planes: top left, Record 311 ('3-in-one'), a Preston design; top right, Norris skew-mouth steel rebate plane; centre, Norris A7 steel adjustable shoulder plane; bottom left, slightly smaller Norris (London pattern) adjustable shoulder plane with non-original stained infill; bottom right, a user-made bullnose shoulder of the Spiers type. (Christie's South Kensington)   (**P**)

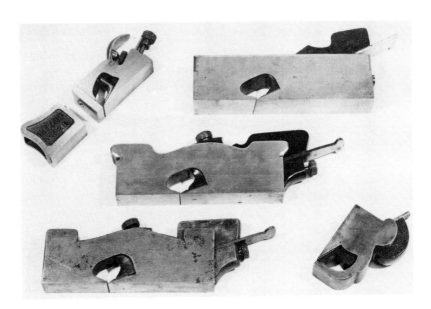

Two other types of British metal plane are the chariot and thumb planes. The latter are not often seen, although they are really the equivalent of the ever popular American block planes. The title 'thumb' planes seems inappropriate for these planes at five inches long – larger than most wood thumb planes or the metal chariot planes. Buck and Ryan catalogues in the 1930s described Norris thumb-planes as 'favoured by pianoforte case makers', suggesting that they had little appeal outside this rather specialized trade – hence, perhaps, their rarity. In the 1930s, these and shoulder, chariot and bullnose planes were available with the Patent Adjustment originally fitted to bench planes only. However, recent prices at Christie's South Kensington of even the commonest of these adjustable special purpose planes, the shoulder, are an indication of the small numbers in which they originally sold.

Chariot planes by known makers form a small proportion of those extant; possibly their small size made them particularly suitable for home manufacture. Many also probably date from the third quarter of the nineteenth century or earlier, when a greater proportion of metal planes were home-made or home-finished in any case. Many, perhaps the majority, are in brass or bronze rather than iron, and the normally very short front sole is often formed as a separate, L-shaped piece attached to the toe by two screws. There is no visible infill, but the bulbous wedge forms a hand-grip and is often of ebony. Most chariot planes are only three or four inches long, usually with the cutter in the 'bullnose' position as close to the toe as possible, but the less common 'Irish' pattern was about nine inches long, and more like an American block plane in appearance, although retaining the bulbous wedge of the English chariot plane. Of English makers, only Preston appear to have produced these on a regular basis, although they are related perhaps to the less elegant low-angle block planes often found in pattern-makers' tool kits.

Production of British metal planes declined after the 1914–18 war: Spiers faded from the scene about 1930, and fewer good quality user-made planes survive from this period. Norris carried on into the 1940s, and was revived after the war as Norris Planes and Tools Ltd.; planes were made by this firm until 1952. These late planes were not dovetailed, but made from solid channel-section rolled steel (how this was achieved in the case of coffin-sided planes has yet to be explained to me). The sides were often finished with a pattern of rotary lapping.

Probably the last of the metal plane-makers to survive was Arthur Price of London, who carried on into the 1960s (one of his shoulder planes is illustrated on page 46). More recently (since 1976), the Henley Optical Company has been making a range of adjustable metal planes of the traditional types, although not everyone agrees that these are superior to the Norris products. They are, inevitably, very expensive.

In general, planes by known makers are more attractive to

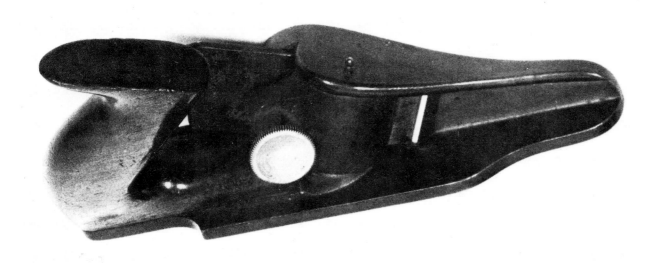

55
Lancashire pattern rebate plane: made of bronze, with a mahogany handle projecting at approximately 45 degrees to the sole, this can be used for trimming either the base or side of a rebate. This example was made by an apprentice in Liverpool in the early years of this century. (Christie's South Kensington) (P)

collectors than those which are anonymous or obviously made by their original owners. However, some of these user-made planes are of superb quality, and they offer greater variety for the collector than can be found in the products of Norris, Spiers, Mathieson, Buck, Holland and the others – and there is variety enough in those, too. Many tool dealers offered castings and other metal parts for owners to complete themselves, and the castings were often made in small or 'one-off' quantities in local foundries. To confuse the issue, Norris gunmetal levers were offered in many tool catalogues, so that some home-made planes may have Norris levers. However, I cannot recall having seen a Norris lever on a plane that was not obviously by Norris, and it is likely that Norris levers so supplied were not marked with the Norris name. That they were illustrated in catalogues with the name is no evidence on this point, since it was common practice (with Norris and many other advertisers) to include their name on tools in catalogue illustrations where it never appeared on the tools themselves.

A characteristic feature of Norris was their cutting irons, which were ground all over and stamped with a simple 'Norris London' logo after this machining, sometimes with the maunufacturer's name, Sorby, on the reverse. They were, like the cutters in nearly all British metal planes, of parallel thickness throughout, in order to preserve the fine mouth throughout the life of the cutter. These cutters were also offered separately. One further Norris plane must be noted, although it is not strictly a metal plane, since the stock is of wood; this was made in the 1920s and 1930s in smoothing, jack and try sizes and consisted simply of a Norris iron with lever cap and Patent Adjustment fitted in a traditional beechwood stock. The idea of fitting an adjustment mechanism to a wooden stock was not new; as early as

the 1850s, Fenn's Patent smoothing plane had incorporated a screw vertical adjustment, and wood versions of the Stanley Bailey planes and their imitators were made for many years in the USA.

A highly eccentric metal combination plane was on sale in the same inter-war period, but not from Norris; this was the Howkins Combination plane, a small tool made in various models (A to E), designed to cut grooves where no ordinary plough could go – round curves, up to stopped ends, at angles, with dovetailed sides or even with right-angled undercutting. To cope with cutting across the grain, two cutters were clamped together, the lower one having two projecting spurs. The cutters were mounted in a unit which was adjusted in a vertical plane (i.e. at right angles to the sole) rather like the slide-rest on an engineer's lathe.

Although more conventional combination planes are usually American in origin, this chapter should not close without mention of a much earlier English model: Silcock's metal plough, patented in 1844, was surprisingly similar in concept to the familiar American types, but pre-dated most of them by some twenty years.

56
Two compromises: *left*, a wood smoother by Kimberley with cast-iron mouth (*c.* 1902); *right*, a Norris wood jack plane with the Patent Adjustment (*c.* 1937).
(Christie's South Kensington)   (P)

3 AMERICAN-TYPE METAL PLANES

In Great Britain in the nineteenth century, metal planes were developed as expensive tools to meet the requirements of high-class joiners and cabinet-makers. They never replaced ordinary wood planes, and were not expected to. In America, events took quite a different turn. There, metal was adopted as a more suitable material than wood for the improvements in ease of use devised by various inventors; it also provided a more stable material than wood, with less liability to warp and to wear, and was better suited to mass production methods.

The ancestry of the American metal plane is generally traced back to a simple cast iron plane patented by one Hazard Knowles of Connecticut in 1827. This had two features now familier to users of Bailey bench planes: the flat iron casting with sides tapering to front and rear from a high point at the throat, and the turned wood front knob. The latter was a compromise between the horn-shaped fore-grips of many continental planes and the much flatter, square 'buns' on British panel and other planes.

The Knowles plane had no form of adjustment, and the cutter was held in place by a wedge housed in tapered slots cast in the sides of the throat. The rear handle was of the closed type on larger models, and sat in a recess cast in the stock. There was no wood infill. The smoothing plane had tall, circular handles at front and rear, both again housed in raised bosses in the casting. Apart from overcoming the problems of wear and warping that can occur with a wooden stock, this design also made for a plane with a very low centre of gravity, giving the user a better 'feel' for the work, and making the plane easier to hold square to the work when planing narrow edges.

The subsequent history of the development of adjustable planes in the USA has been so well covered by Roger K. Smith's recent book *Patented Transitional and Metallic Planes in America, 1827–1927* that there seems no need to repeat it here. In any case, this is very much a subject for American collectors only, as planes of the transitional type very seldom come the way of collectors elsewhere. Briefly, the train of

events was that after various attempts by several inventors in the 1840s and 1850s to provide screw adjustment for cutters in wood planes, Leonard Bailey of Massachusetts in 1855 patented a metal scraper plane in which the cutter was adjustable by a thumb-screw. He then proceeded to adapt the idea to suit a plane with a conventional cutter, which he achieved by fixing the cutter assembly to the upper, hinged section of a horizontally divided stock. A thumbscrew adjusted the position of this hinged section and hence the height of the cutter.

There followed a simplified version in which the stock was made in one piece but the cutter was attached to a rocking 'seat'. From this developed the fixed seat or 'frog' found on Bailey-type planes to this day. This was in 1867. The seat no longer had to rock because the cutter itself moved up and down, controlled by a lever with a forked end engaging a groove in the adjusting knob. Although the patent showed a horizontal knob similar to that of the rocking seat models, surviving examples so far reported have all had the knob in the more familiar vertical position, with a horizontal axis. The basic form of the

57
Early and late Bailey planes: the 4½ (foreground) has no lateral adjusting lever and must date from 1884 (the year the 4½ was introduced); the lateral lever came in 1885. Notice also the absence of a frog adjusting screw, and the more pronounced forward slope of the handle. The 5 dates from the immediate post-World War II period.

adjustable bench plane as we now know it in the English-speaking world was thus established, although a wood-based version was also made. This had the mechanism mounted on a cast-iron bed which was screwed to a thick wood sole, and planes of this type continued to be offered by Stanley and other companies in America for many years.

Another feature patented by Bailey was a thin, parallel cutter. The parallel thickness was an essential feature of a design in which the cutter slid up and down while retained in position, and the thinness saved metal and also labour in grinding and sharpening. A further aid to the adjusting system was the cam-operated lever cap. This provided constant pressure on the cutter once the central retaining screw was set, unlike the screw lever which had to be re-set at each sharpening.

Bailey planes of this early period are few and far between, but even when they do surface, they are easily mistaken for later models at first glance. Apart from the absence of the lateral adjustment lever (not patented until 1884) the only obvious distinguishing feature is the flat face of the adjusting knob, stamped with patent dates and, usually, Bailey's name. The more familiar recessed knob appeared in 1874 and, initially, continued to be stamped in the recess with Bailey's name. From 1869, the planes were manufactured by the Stanley Rule and Level Company of New Britain, Connecticut. In the ensuing years, there were various agreements (and disagreements) between Bailey and Stanley, and in 1875 Bailey established a new business manufacturing 'Victor' planes. These had a new form of adjustment, using a flat knob with a spiral groove on its face. This led to further problems with Stanley, whose Justus A. Traut claimed prior invention of the scroll device, and eventually Stanley bought the Victor plane business in 1884. Only one model, the No. 20 circular plane, survived for more than a few years in production, and this was later fitted with the familiar original Bailey type of adjustment, with a cam lever in place of the Victor's screw type. Although Leonard Bailey lost contact with the manufacture of his planes, the Stanley company

58
Palmer's Metallic Plane, *c.* 1870: this battered example of a rare early competitor of the Bailey adjustable plane turned up at Christie's South Kensington in 1980. Made by the Metallic Plane Company of Auburn, New York, it had a corrugated sole (patented in 1869) and no less than three adjustment levers.
(Christie's South Kensington)   (P)

59
Stanley No. 1: the smallest of the Bailey bench planes, at only 5½ inches long, this miniature model is seldom seen, although it was offered from 1870 to 1943.
(Christie's South Kensington)   (P)

later (in the early 1900s) incorporated his name into the casting of the stock of most bench planes, where it remains to this day as a memorial to a man who, more than anyone else, revolutionized plane design.

Justus A. Traut's scroll adjustment was never put into production by Stanley, but his 1884 lateral adjustment lever was. In 1888, it was improved by the addition of a disc to reduce friction at the point of engagement with the slot in the cutter. Now, the Stanley-Bailey plane as we know it was established; subsequent changes included a left-hand thread for the adjustment knob (so that it screws *in* to lower the iron, which feels more 'natural' to most users) in the 1890s, alterations in the frog (most importantly a raised seating patented in 1902) and an adjusting screw for it patented in 1907, a taller front knob and the Stanley name cast in the lever cap (1925). Handles were mostly of rosewood until the Second World War, and in some cases afterwards. By this time, there was a Stanley factory in Sheffield, the centre of the English edge-tool industry, and the English planes mostly have stained beech handles except for the most recent, which are equipped with rosewood-coloured plastic. Another recent change in the UK which I have noted with regret is the replacement of the traditional yellow cardboard boxes with dark green labels (in which Stanley tools have been packaged for many years) with a gaudy pictorial design no doubt intended to appeal to customers in self-service shops. The English planes, however, continue to have black-painted castings, while those in America have changed to blue and then red.

For the enthusiastic collector of Stanley planes, there are numerous minor changes which are useful dating points, and these are set out in Roger Smith's book mentioned earlier, and also Alvin Sellens' *The Stanley Plane*, an excellent reference work in which it is easy to find details of any Stanley plane by reference to its model number.

Many tool collectors find little to attract them in these mass-produced iron planes, but they do command a strong following, particularly in their native land, and it is perhaps the very fact of their being early examples of mass-production that makes them interesting. Furthermore, they are eminently usable tools – even those who prefer to use nothing new in their workshop can enjoy the benefits of this 'modern' tool with an example that is nearly a century old. Earlier models are preferable to current versions, in my view, partly because of their rosewood handles (provided that the projection at the top of the rear handle has not broken off, as it often has, leaving a sharp edge to dig into the hand) and partly because, after years of use, they are smoother and more comfortable to hold. Cutters wear, of course, but a modern cutter can be purchased at any High Street tool-shop and will fit an early plane without any difficulty. On the subject of cutters, the company name stamp at the top is always worth inspecting, as it changed in style from time to time and is a dating feature, although always with the proviso that cutters are easily replaced or interchanged between planes of similar width, and so are not reliable

60
Stanley frogs, old and new styles (from a 1928
Stanley catalogue). The new-style Bailey frog
(without the adjusting screw) was patented in 1902.

172   S T A N L E Y   T O O L S

FROGS FOR "BAILEY" AND "BED ROCK" PLANES

From time to time improvements have been made in both the "Bailey" and
"Bed Rock" Iron Planes, which necessitated changes in the construction of the
Bottom and Frog, making it impossible to use the new style Frog in an Old Style
Bottom, or the Old Style Frog in a New Style Bottom.

TO INSURE YOUR ORDER FOR FROGS BEING CORRECTLY FILLED,
ALWAYS STATE WHICH STYLE PLANE YOU HAVE.

BAILEY OLD STYLE   BAILEY NEW STYLE

For a time an intermediate style was made having same Frog and Bottom as the
latest design, except that there was no Frog adjusting screw, consequently no clip
on the Frog.

The latest design Frog or Bottom will be furnished for both the intermediate
and new style Planes. If your plane is of the intermediate pattern, remove the
steel clip from the Frog and the parts will fit.

The difference in construction of the Frogs and Bottoms in the "Bailey" Planes
is shown in the illustration above.

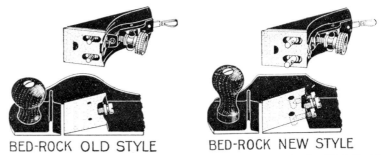

BED-ROCK OLD STYLE   BED-ROCK NEW STYLE

The difference in construction of the Frogs and Bottoms in the "Bed Rock"
Planes is shown in the illustrations above.

by themselves as proof of a plane's date. The most notable of these
changes was the takeover of the Stanley Rule and Level Company by
the separate Stanley Works in 1920; the old company was then known
for some years as the Stanley Rule and Level Plant, and products often
bore an SW heart-shaped logo under the Stanley name. Until the
Second World War, cutters on both Record and Stanley planes were
of the traditional composite hard and soft steel construction.

The Stanley range of planes is known by its numbers; these start with the Bailey bench planes, Nos. 1–8. No. 1, the smallest, is a mere $5\frac{1}{2}$ inches long, and has all the charm of the miniature; it lacks the lateral adjustment regardless of date, but is otherwise similar in design to the full-size models. Although it was made into the 1940s, it is rarely seen and commands a high price when examples do come on the market — so much so that modern copies have been reported. No. 2, at 7 inches, is also not found with great frequency, but No. 3 (8 inches, with a cutter width of $1\frac{5}{8}$ inches) is still in production. No. 4 is the most popular and useful of the smoothers; it is 9 inches long, with a 2-inch cutter. The No. $4\frac{1}{2}$ was not introduced until 1884, and was the largest smoother at 10 inches by $2\frac{3}{8}$ inches. The extra size is often unnecessary, and in my own view this is a less comfortable plane to use than the No. 4. It went out of production in the States at the end of the 1950s, although it is still made in England by both Stanley and Record.

Jack planes are the Nos. 5 and $5\frac{1}{2}$, at 14×2 inches and $15 \times 2\frac{1}{4}$ or $2\frac{3}{8}$ inches respectively; the No. 6 is described in the Stanley catalogues by the archaic term (in English usage) 'fore' plane. It is the same width as the $5\frac{1}{2}$, but 18 inches long. Nos. 7 and 8 are jointers at $22 \times 2\frac{3}{8}$ and $24 \times 2\frac{5}{8}$ inches – the equivalent of wood try-planes. The No. 8 is no longer made by Stanley, although the Record 08 survived until 1982.

So far, so good. Henceforth, the numbers become many and varied and I will not test the reader's concentration by attempting to explain every one. Rather, I will look at the more important types of plane, and for a number-by-number account I recommend once again the Sellens volume mentioned above.

The main variation on the standard Bailey plane was the Bed Rock. This had a modified seating for the frog, and was identifiable, in the case of post-1914 examples, by a flattened top to the hump in the sides. The frog was equipped with a screw fine adjustment, which was later (1914) adopted also on standard Bailey planes. At about this time, Bed Rocks acquired a further refinement, in which the screws attaching the frog to the bed were replaced by pins clamped by screws at the rear; this enabled adjustments to the frog to be made without removing the cutter, a 'look, no hands!' trick that smacks of trying rather too hard to keep the Bed Rock 'up-market' of the Bailey at all costs. Bed Rock planes were made from 1902 until the 1930s and, in the case of one or two sizes, the early 1940s. Like the standard planes, Bed Rocks were available with grooved or 'corrugated' soles, planes so equipped being identified by a 'C' suffix to the model number. All the Bed Rock numbers were prefixed by 60, the final digit indicating the size as with the Bailey models; the numbers ran from 602 to 608.

A very large and almost absurdly diverse range was that of block planes – small, single-iron planes designed for use in one hand and suited to trimming end-grain, or for general purposes by amateurs not wishing to go to the expense of a fully fledged Bailey plane.

The earliest block plane, the No. 9, was a cabinet-maker's tool and

62
(*Right*) Stanley block planes: from the 1930 Buck and Ryan catalogue.

61
Stanley No. 9 cabinet-maker's block plane: the first of the Stanley block planes (introduced in 1870), this is clearly based on the European mitre planes. (Christie's South Kensington)   (P)

Buck & Ryan
Stanley Block Planes

K779. No. 100. 3½ in. long, 1 in. cutter ... ... **2/-** each.

K780. No. 101. 3½ in. long, 1 in. cutter ... ... **1/6** each.

K781. No. 102. 5½ in. long, 1⅜ in. cutter ... ... **3/-** each.

K782. No. 103. 5½ in. long, 1⅜ in. cutter with lever adjustment ... ... **4/3** each.

K783. No. 110. 7 in. long, 1⅝ in. cutter ... ... **4/3** each.

K784. No. 120. 7 in. long, 1⅝ in. cutter, lever adjustment ... ... **6/-** each.

K785. No. 203. 5½ in. long, 1⅜ in. cutter, adjustable. Designed for manual training use ... ... **5/6** each.

Double-end Plane

K786. No. 130. 8 in. long, 1⅝ in. cutter ... ... **6/3** each.

Has two slots and two cutter seats. The centre seat and slot to be used for ordinary block plane work, the other slot and seat for use when it is desired to work as a bull-nose plane.

Double-end Adjustable Plane

Combination block and bullnose plane with two slots and movable cutter seat. For use as a bullnose plane, reverse the cutter seat by throwing over the adjusting wheel.

K787. No. 131. 8 in. long, 1⅝ in. cutter ... ... **10/-** each.

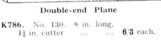

K788. No. 220. 7 in. long, 1⅝ in. cutter, with screw adjustment ... ... **6/3** each.

Block and Rabbet Plane

A detachable side will easily change it from a block plane to a rabbet plane. The cutter is adjustable and set on a skew which cuts across the grain while working straight.

K789. No. 140. 7 in. long, 1¾ in. cutter ... ... **10/9** each.

Fitted with a knuckle-jointed lever which, being entirely of steel, is practically indestructible and locks the cutter.

K790. No. 18. 6 in. long, 1⅝ in. cutter ... ... **10/9** each.

K790A. No. S18. All-steel 6 in. long, 1⅝ in. cutter ... **13/3** ,,

K790B. No. 19. 7 in. long, 1⅝ in. cutter ... ... **11/6** ,,

K791

The four following planes are designed to be easily held in one hand. Cutters adjustable. Mouths adjustable for coarse or fine work.

K791. No. 9½. 6 in. long, 1⅝ in. cutter ... ... **8/-** each.

K791A. No. 15. 7 in. long, 1⅝ in. cutter, japanned ... **8/9** each.

K791B. No. 16. 6 in. long, 1⅝ in. cutter ... ... **9/9** each.

K791C. No. 17. 7 in. long, 1⅝ in. cutter, nickel-plated **10/6** each.

For **Stanley Plane Irons and Fittings.**

See page 85.

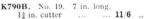

Low Angle Plane

Mouths adjustable for coarse and fine work

K792. No. 60½. 6 in. long, 1⅜ in. cutter ... ... **9/-** each.

K792A. No. 65½. 7 in. long, 1⅝ in. cutter ... ... **9/6** ,,

was very clearly derived from the English iron mitre planes. It had a Bailey-type cutter adjustment (which necessitated the fitting of a double iron), a spherical hand-grip on an extension at the rear and a metal handle which could be fitted to either side of the throat, straddling the top of the side-plate. This is one of the rarer Stanley planes, although it ran from 1870 to 1943. The more typical block planes are mostly six or seven inches in length (apart from some models of 'thumb' size) with shaped sides like the Bailey planes and cutters held by a combined lever-cap and hand-grip. This cap is variously fastened against the cutter by a knurled handwheel, a lever-operated snail cam or an ingenious articulated knuckle-joint mechanism in the cap itself. With the latter, the upper part of the cap is lifted to release the cutter. Adjustment for the cutter was provided on many block planes, either by a variation on the original, vertical-axis Bailey patent or by a 'long-screw' device axially parallel with the cutter. A lever adjustment was also offered at one time, and a lateral adjustment lever usually accompanied the Bailey-type adjustment.

63
Anglo-American planes: GTL bronze smoother (*rear*) and Record 018 adjustable block plane with quick-release lever-cap.

64
Stanley 72½ chamfer plane with beading
attachment: the basic chamfer plane (without the
beading attachment and its cutters) was No. 72.
(Christie's South Kensington)   (P)

One of the more ingenious block planes had a skew-iron extending
to the extreme right-hand edge, Badger fashion, with a separate
sideplate which could be attached for use in non-rebate work. This
was No. 140. From the late 1890s many block planes had recesses
ground in the sides of the stock to form hand-grips; this was referred
to as the 'Hand-y Feature'. Another piece of ingenuity was an
adjustable mouth, formed not by a movable frog but by a sliding
section in the front sole, fastened by a knurled brass thumbscrew on
top which doubled as a fore-grip. From about 1898 an eccentrically
slotted lever was added below this knob, which made adjustment of
the front sole much easier. Although several of the block planes are
still in production, there are plenty of obsolete and early models to
interest the collector.

Stanley appear to have tried to produce an equivalent to every
existing wood plane; although its greater wearing qualities made iron
an excellent substitute for much-used tools like bench planes, for the
more specialized planes which received little use, iron seems
unnecessarily durable (notwithstanding its brittleness) and indeed
too intractable for such purposes. Moulding planes as such were not
made; the nearest that Stanley came to individual moulding planes
were the corner-rounding plane and the chamfer plane. The latter
(Nos. 72 and 72½) was another example of Justus A. Traut's fertile
mind; the main sole was V-shaped, but the front section was flat and
vertically adjustable on the line of the cutter angle. An extra fitting
was available (it made the 72 into a 72½) which took the place of the
normal front section and carried moulding cutters of the type fitted to
the 66 hand-beader. However, as these are scrapers rather than plane
cutters, they make the plane into a form of scratchstock rather than a
moulding plane. In chamfer form, the plane could be fitted with a

bullnose front, provided on both the 72 and the 72½ from 1909.

For forming rebates (or, as they are usually known in America, rabbets) there were several planes, some of which suggest some influence from the British metal planes. An early (but short-lived) offering was the No. 80 (or, with spurs, No. 90) wood-filled metal rabbet plane. There was no dovetailing, though – a bent-up steel case was used, and the slotted skew cutter was held in place by a clamping screw which passed through the rear of the stock to a wing nut on the heel. This model was discontinued in 1888, but the No. 90 was later re-allocated to an all-iron bullnose shoulder plane that is still with us. This has an adjustable mouth, achieved by forming the stock in two sections divided horizontally, with the upper section curving down at the front to form the toe; the two parts are attached by a screw in a slot, enabling the upper section to be slid back or forwards so as to open or close the mouth. A larger, non-bullnose version is the No. 92. For a brief period at the end of the 1930s, two bullnose planes without adjustable mouths were made as 90J and 90A, and these must certainly have been inspired by the English Preston 1363/Record 076 plane of similar size and shape.

A cheaper bullnose plane (also still to be seen in the tool-shops) was the 75; this had no cutter adjustment, but did use the same system of mouth adjustment as the 90. It was the only one of this type to be copied by Record, who had their own forms of mouth adjustment; unusually, it is the Record version (075) which has gone out of production rather than the Stanley.

For ordinary rebating work, the 78 was the most likely tool to be used. It was (and is) a fillister, and could be used with the cutter in a central or bullnose position; hence its title 'Duplex Fillister'. By removing the fence and depth-stop, it could be used as a simple rebate plane. The 78 was introduced in 1884 and has always had a simple screw lever cap, although a somewhat crude lever adjustment has been fitted since 1925. Early models (pre-1909) are more elegant than their successors, with a lower main stock and floral decoration on the handle. A non-duplex, non-fenced version was made in three sizes as the 180–182 and 190–192 (the latter being fitted with spurs). These models appear to have been less popular than the more versatile 78. The current Record version is an improved model, with the fence supported on two arms instead of one.

A skew-cutter version was offered by Stanley as the 289 (this had both fence and depth-stop) and an alternative form of duplex fillister was the 278, which was converted for bullnose work by removing the toe rather than by moving the cutter. It had a very low cutter angle (with lever adjustment) and was of different (and rather inelegant) design to the 78 and its derivatives. Perhaps the most improbable-looking of all the rebate planes is the No. 196, which is unlikely to come the way of many collectors. This was capable, according to the catalogues, of cutting rebates on the outside or inside of curved or

65
Stanley special purpose planes (1930).

Buck & Ryan

Stanley Planes—*continued*

K793. No. 75. Bullnose Rabbet Plane. 4 in. long, 1 in. cutter, adjustable mouth ... ... **2/6** each.

K796

K796A

Side Rabbet Planes

For side-rabbeting in trimming dados, mouldings and grooves of all sorts. A reversible nosepiece gives the tool a form whereby it will work close up into corners. Fitted with depth gauge.

K796. No. 98. 4 in. long, ½ in. cutter, right hand, nickel-plated **8/-** each.
K796A. No. 99. 4 in. long, ½ in. cutter, left hand, nickel-plated **8/-** each.

Double-End Match Planes

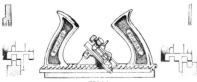

K800

These planes have two separate cutters, a plow and a tongue tool, both governed by one permanent fence. Both tongue and groove are cut by working the tool in the same direction by merely reversing it end for end.

K800. No. 146. Cuts ⅛ in. groove on boards ⅜ in. to ½ in. Centres on ⅜ in. Nickel-plated ... ... **15/-** each.
K800A. No. 147. Cuts 3/16 in. groove on boards ½ in. to ¾ in. Centres on ⅝ in. Nickel-plated ... ... **16/-** ,,
K800B. No. 148. Cuts ¼ in. groove on boards ½ in. to 1 in. Centres on ⅞ in. Nickel-plated ... ... **17/-** ,,

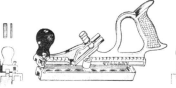

K802. Stanley Matching Planes

To cut a tongue on the edge of one board to fit a groove in the edge of another. The straightness of both tongue and groove and their distance from the surface, is governed by a fence. The swinging Fence Match Plane has two plow cutters of the same width, and one extra wide cutter. The fence in one setting exposes two cutters for cutting the tongue, and when reversed leaves only one exposed for cutting the groove.

K802. No. 48. Cuts ⅝ in. groove on boards ¾ in. to 1¼ in. Centres on ⅞ in. Nickel-plated ... ... **20/-** each.
K802A. No. 49. Cuts 3/16 in. groove on boards ½ in. to ¾ in. Centres on ½ in. Nickel-plated ... ... **19/-** ,,

K794. No. 90. Bullnose, 4 in. long, 1 in. cutter, adjustable mouth. **14/3** each.

Side Rabbet Plane

A convenient plane for side-rabbeting, in trimming dados, mouldings and grooves of all sorts. A reversible nosepiece allows it to be worked up into close corners when required for left or right hand.

K797. No. 79. 5½ in. long, ½ in. cutters ... ... **8/-** each.

Cabinet-makers' Rabbet Planes

For fine work of extreme accuracy. Both sides are square with the bottom, and sides and bottoms are machine ground. Can be worked either right or left hand. The width of the throat is adjustable. Cutters are adjustable. By removing the front a chisel plane is obtained.

K795. No. 92. 5½ in. long, ¾ in. cutter, nickel-plated ... ... **14/3** each.
K795A. No. 93. 6½ in. long, 1 in. cutter, nickel-plated ... ... **17/-** each.
K795B. No. 94. 7½ in. long, 1¼ in. cutter, nickel-plated ... **19/6** each.

Edge-Trimming Block Plane

For trimming or smoothing edges for a close fit. Cutter works on a skew. Wood blocks of various bevels may be attached, enabling the user to make a slanting cut.

K798. No. 95. 6 in. long, ⅞ in. cutter ... ... **6/9** each.

K799. Iron Rabbet Planes

These planes lie flat on both sides, and can be used with either right or left hand while planing into corners. Fitted with spur and detachable depth gauge.

K799. No. 190. 8 in. long, 1½ in. cutter ... ... **10/9** each.
K799A. No. 191. 8 in. long, 1¼ in. cutter ... ... **9/9** ,,
K799B. No. 192. 8 in. long, 1 in. cutter ... ... **9/3** ,,

K801

Duplex Rabbet and Fillister Planes

They have two seats for the cutter, one for regular and the other for bullnose work. Also a spur and a removable depth gauge. The adjustable fence can be used on either side, and slides under the bottom for regulating width of the cut. The rear cutter is adjustable endwise.

K801. No. 78. 8½ in. long, 1½ in. cutter ... ... **11/9** each.

straight edges. I have never handled one of these, but suspect that a considerable degree of dexterity would be needed to make it perform satisfactorily.

From rebates one turns naturally to ploughs, and it was here that Stanley excelled themselves. The original development in this field was largely the work of C. G. Miller and our old friend Justus A. Traut. The very early models (which do come to light from time to time, even in Great Britain) have a richly ornamented cast iron or bronze stock which extends upwards at the rear to support the rosewood handle at top and bottom. For fillister work, a wide sole was available, and with this a special fence designed to clear it was needed. Perhaps the most interesting feature of the Miller's Plow was its retention of the traditional thick plough cutter, located on the skate by a V-groove. This was held in its housing in the stock by a pivoted lever clamped by a knurled thumbscrew on top. The design was patented in 1870, and the Miller's Patent Adjustable Metallic Plow was covered by catalogue Nos. 41–44, the four permutations depending on the use of

66
Stanley No. 41 Miller's Plow: introduced in 1870, but this is a later version, as indicated by the presence of a slitting cutter and the shape of the fence, which has two pairs of holes for the arms and curves outwards to clear the fillister bed (here shown detached). The extra holes were added in 1887.
(Christie's South Kensington)   (P)

67
Stanley 46 skew-cutter plough and fillister: this example of the 46 (introduced in 1874) dates from around 1909. The full set of cutters contains ten, plus a slitting cutter.

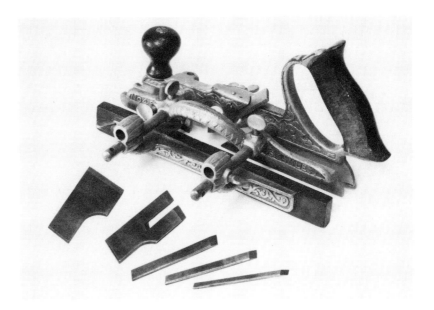

iron or bronze and the presence or otherwise of the fillister bed and appropriate fence.

A bullnose version of the Miller's Plow was later added, and this, shorn of its ornament, remained in the catalogues up to the 1940s. The original models, however, had been superseded by the early 1890s by considerably revised designs developed from Dorn's Patent Combination Iron Dado Plane of 1872. The first of these was the skew-cutter 46 (attributed in contemporary advertising to Traut) and from this evolved the first Stanley combination plane designed to work mouldings rather than grooves and rebates – the No. 50. The cutters in this model were clamped to the main stock by a second stock which slid on the fence-arms and provided a second skate aligning automatically with the edge of the cutter. Thus both quirks of the bead were supported, although one was inoperative if a side-bead was being worked, with the fence aligned on the edge of the bead itself. The handle was of metal, although most of the post-Second World War 50s had wood handles. It soon became clear that the 50 was just as suitable for grooving as for beading, and by 1936 it came with 17 cutters, including tongue-and-groove, fillister and plough irons.

This and most of the succeeding combination planes were largely the work of Justus A. Traut, whose next offering was the 45. This came in 1885; it was larger than the 50 but retained the latter's basic shape, with a wooden handle and floral ornament in the castings inherited from the Miller ploughs and the 46. Also recalling the 46 was a turned fore-grip on the front of the main stock (later moved to the front fence arm on the 45) and at the rear of the stock was a slitting cutter, a double-edged knife tool intended as a means of cutting thin

RICHARD MELHUISH LIMITED, LONDON, E.C.

STANLEY PLANES

No. 100W.

Price complete with 23 cutters ... ... **£2 : 16 : 9**

STANLEY "FORTY-FIVE" PLANE

Seven tools in one in compact and practical form. *1.* Beading and Centre Beading Plane. *2.* Plough. *3.* Dado. *4.* Rebate and Fillister. *5.* Match Plane. *6.* Sash Plane. *7.* Slitting Plane.

The Plane is fitted with spurs for use across the grain, and can be used either right or left hand.

With each Plane are furnished twenty-three cutters, all of which are shown. The cutters, together with the Plane, are packed in a neat, substantial box.

All metal parts are nickel-plated. The handle, knob, and adjusting fence are made of selected Rosewood.

68

Stanley 45: dealers' catalogues often used out-of-date blocks for their illustrations and this one (Melhuish, 1925) shows the current model of the '45' on the left and an earlier, more ornate, version on the right.

timber more quickly than it could be sawn by hand with a rip-saw. The sliding stock could be fixed at any point across the width of the cutter, since the latter was clamped independently to the main stock, an important distinction from the 50, on which the sliding stock itself acted as a clamp and always aligned automatically with the extreme edge of the cutter.

Cutters offered with the 45 numbered eighteen initially, increasing to twenty-three by 1925; they included beads, plough, fillister, stick-and-rabbet sash and tongue-and-groove irons. A range of optional extra cutters included reeds, flutes and hollows and rounds; for the latter to work satisfactorily, more support was needed than was afforded by the two skates, and the hollow and round cutters came with appropriately shaped full-width metal soles. The 45 was perhaps the best all-round tool of the combination planes, performing most of the tasks likely to be needed by a user not equipped with wood moulding planes, and lacking the complication of the 55. It was made until 1962, and the Record 405 Multi-Plane (which is virtually identical) remained in production until 1982.

The Stanley 55 Universal plane first appeared in the mid-1890s and was a logical elaboration of the 45. The 55 had two fences with pivoted faces (for working on a slant) and one also had a fine screw adjustment. Its most important feature was the vertical adjustment of the skate on the sliding stock; this enabled asymmetric mouldings such as ogees, coves and ovolos to be worked. The problem of intermediate support was overcome not by full-width soles as on the 45, but by a rather inadequate short, removable skate in front of the cutter, adjustable vertically and laterally and provided with an optional 'shoe' to give a wider working surface when needed. Even with this 'Auxiliary Center Bottom', as it was charmingly described in

69

Stanley 55: also from the 1925 Melhuish catalogue.

TOOL AND MACHINE MERCHANTS

UNIVERSAL COMBINATION PLANE

A THOROUGHLY GOOD TOOL EXTENSIVELY ADOPTED BY PRACTICAL WORKMEN

No. 191W.

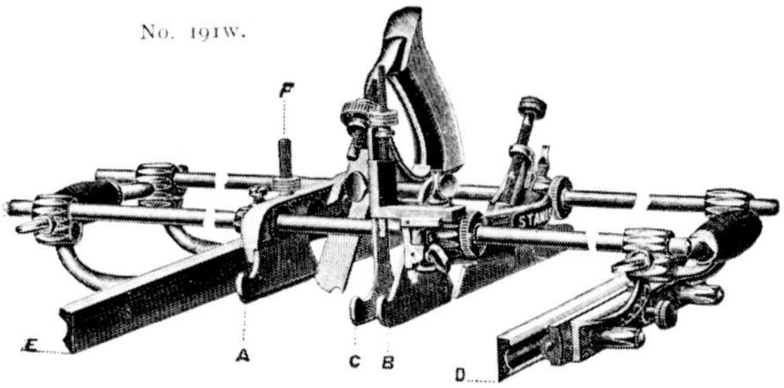

STANLEY "FIFTY-FIVE" PLANE

This Tool, in addition to being a beading and centre beading plane, a plough, dado, rebate, fillister, and match plane, a sash plane and a slitting plane, is also a superior moulding plane, and will accommodate cutters of almost any shape and size. In fact, it is "A PLANING MILL WITHIN ITSELF."

The regular equipment sent with the Plane comprises fifty-five cutters, all of which are shown. These cutters, together with the Plane and all its attachments, are packed in a neat, substantial box.

A further line of forty-one cutters are carried in stock. Cutters of practically any form can be used in the Plane, which the owner can make from blanks or order from sketch.

All metal parts of the Plane are nickel-plated. The handle and fences are made of selected Rosewood, and every part is well finished.

This Plane consists of :

A MAIN STOCK (A) with two sets of transverse sliding arms, a depth gauge (F) adjusted by a screw, and a slitting cutter with stop.

A SLIDING SECTION (B) with a patent vertically adjustable bottom.

THE AUXILIARY CENTRE BOTTOM (C) is to be placed in front of the cutter, as an extra support or stop, when needed. This bottom is adjustable both vertically and laterally.

FENCES D AND E.—Fence D has a lateral adjustment by means of a screw, for extra fine work. The fences can be used on either side of the Plane, and the rosewood guides can be tilted to any desired angle, up to 45°, by loosening the screws on the face. Fence E can be reversed for centre beading wide boards.

AN ADJUSTABLE STOP to be used in beading the edges of matched boards is inserted on the left-hand side of sliding section (B).

By means of the patent adjustable bottom and the auxiliary centre bottom it is possible to use a cutter of practically any shape with this Plane.

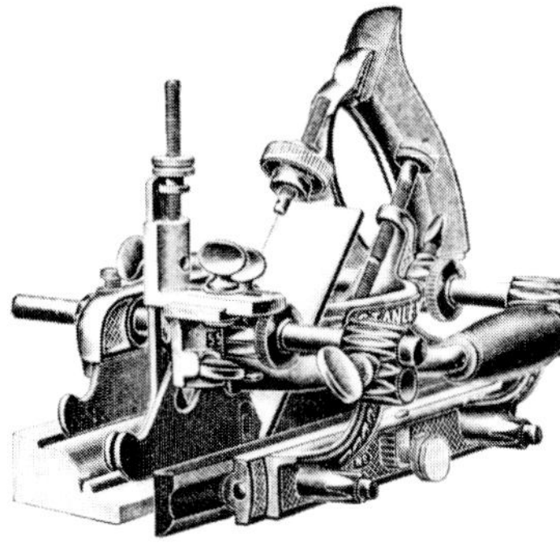

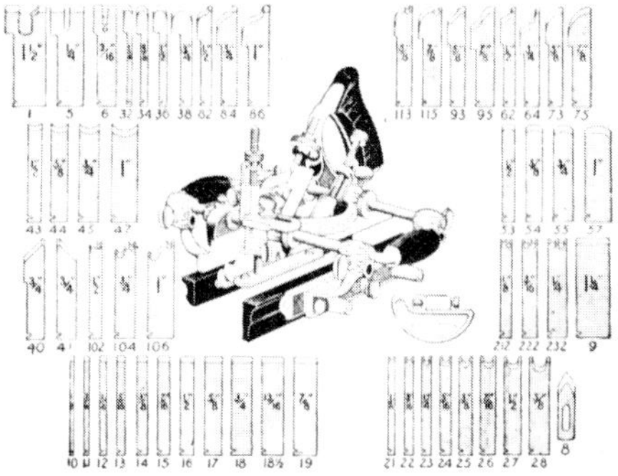

CUTTERS REGULARLY SUPPLIED WITH PLANE

The following Cutters are furnished with each Plane. The prices are given in case duplicates should be required :

No.		Each.	No.		Each.	No.		Each.
1.	1½ in. Sash Tool	4/3	25.	⅜ in. Beading Tool	1/9	62.	½ in. Quarter Hollow ...	3/9
5.	¼ „ Match Tool	4/3	26.	7/16 „ „	2/-	64.	¾ „ „ „ ...	4/3
8.	Slitting Tool	2/6	27.	½ „ „	2/-	73.	⅝ „ „ Round ...	3/9
9.	Fillister	2/-	28.	⅝ „ „	2/6	75.	⅞ „ „ „ ...	4/3
10.	⅛ in. Plough and Dado Tool	1/3	29.	¾ „ „	2/6	82.	½ „ Reverse Ogee... ...	3/9
11.	3/16 „ „ „ „	1/3	32.	¼ „ Fluting Tool	2/6	84.	¾ „ „ „	4/3
12.	¼ „ „ „ „	1/3	34.	⅜ „ „	2/6	86.	1 „ „ „	4/3
13.	5/16 „ „ „ „	1/3	36.	½ „ „	2/6	93.	⅝ „ Roman „	3/9
14.	⅜ „ „ „ „	1/9	38.	¾ „ „	2/6	95.	⅞ „ „ „	4/3
15.	7/16 „ „ „ „	1/9	40.	¾ „ R.H. Chamfer	4/3	102.	½ „ Grecian „	3/9
16.	½ „ „ „ „	1/9	41.	¾ „ L.H. „	4/3	104.	¾ „ „ „	4/3
17.	⅝ „ „ „ „	1/9	43.	½ „ Hollow	1/9	106.	1 „ „ „	4/3
18.	¾ „ „ „ „	1/9	44.	⅝ „ „	1/9	113.	⅝ „ ¼ Rd. with Bead ...	3/9
18½.	13/16 „ „ „ „	2/-	45.	¾ „ „	1/9	115.	⅞ „ ¼ „ „ ...	4/3
19.	⅞ „ „ „ „	2/-	47.	1 „ „	1/9	212.	⅜ „ Reeding Tool, 2 Bd....	1/9
21.	⅛ „ Beading Tool	1/3	53.	½ „ Round	1/9	222.	3/16 „ „ 2 „ ...	1/9
22.	3/16 „ „	1/3	54.	⅝ „ „	1/9	232.	¼ „ „ 2 „ ...	1/9
23.	¼ „ „	1/3	55.	¾ „ „	1/9			
24.	5/16 „ „	1/9	57.	1 „ „	1/9			

Price complete ... ... £5 : 11 : 6

Bronze Medal 1884—Gold Medal 1890

the Stanley literature, working mouldings such as nosings on the full width of the timber is not easy with a 55. According to the instructions issued with the plane, "The working of a return bead of a good shape is often a puzzle with the ordinary wooden bead planes. With the '55' the bead is first made on the edge of the board . . . so that a small quirk is left on the face side." In practice, I have found that, if working such a cut with a wooden plane is a puzzle, with the 55 it is one of the world's unsolved mysteries. With only the quirks supported by the skates (and then only on the very edge of the timber) and with no depth-stop, the cutter simply digs into the wood like an uncontrolled chisel.

Originally, the 55 came with fifty-two cutters; from 1925 the number was increased, appropriately, to fifty-five, and there were also a further forty-one cutters available as extras (most of them different sizes of profiles included in the standard set). With this vast number of cutters (many of which could produce more than one moulding – beads, for example, could be used for side or centre beads, torus, astragal or even ovolo sections) the 55 equalled, perhaps, two chestfuls of ordinary wood planes. However, it could fairly be said to suffer from the shortcomings of any jack of all trades, and it is notable

70

Fales Plow: a combination plane patented in 1882 by Amos Fales of Denver, Colorado, and manufactured by Otis Smith of Rockfall, Connecticut. Each cutter (over 80 were available) requires its own shaped sole, formed as two separate castings. (A similar arrangement, with one-piece castings, was used for certain optional cutters on the Stanley 45 and Record 405.)
(*Old Woodworking Tools*)

71
Union wood-bottomed plane: like its Bailey
progenitor, this has a cast-iron bed screwed to a
wood block, and a previous owner has taken
advantage of the construction to convert it to a
pattern-maker's plane with interchangeable curved
soles of different radii. The cutter adjustment
consists of a lever locked in place by two knurled
thumb-nuts.

that many 55s (and 45s too) appear today in little used condition,
many of their cutters never having been sharpened. The 55 was also
discontinued in 1962, and was one of the few well known Stanley
models never copied by Record.

Of the many other American firms offering metal planes in the late
nineteenth and early twentieth centuries, the best known were
Sargent (whose products were offered by many British tool-dealers)
Union and the Ohio Tool Company. Ohio even used the same numbers
to identify their models as those of the Stanley originals, except that,
for copyright reasons, each number was disguised by being prefixed
with an 'O'. The same ploy was adopted in England by Record,
although for some reason the 45 became a '405' rather than an '045'.
Some of these rival planes had minor differences from the Stanley
originals (as did many seldom seen variations produced by Stanley
themselves). Union, for example, produced a range with a screw
lever-cap and a lever cutter adjustment with two locking nuts, and
Sargent and Gage produced 'Auto-set' and 'Self-setting' planes
respectively, in which it was claimed that the cutter could be removed
for sharpening and replaced in the plane without needing to be re-
adjusted. Gage were taken over by Stanley in 1919–20, and the planes,

which had all been of the wood-bottomed variety, continued to be made, and were offered also in all-metal versions, under the Stanley banner.

In England, the Birmingham firm of Edward Preston and Sons produced at least two adjustable smoothing planes of Bailey inspiration, though claiming superiority in the use of malleable rather than ordinary cast iron for the stock, and by 1930 they, like Spiers, were offering a conventional Bailey-type plane. In the early 1930s the Preston business, which was as noted for its rules and levels as for its planes, was taken over by Rabone, who subsequently sold the plane side of the business to C. and J. Hampton Ltd. This firm, which had begun making iron vices and cramps before the first war, entered plane manufacture about 1930, and by 1938 had a comprehensive range of Stanley-type planes, supplemented by several Preston models.

Most of the Preston planes adopted by Record (which were all of the shoulder and bullnose family) were re-numbered, although the charming little Preston 1366 bullnose plane, still with a rosewood wedge, retained its old number. Adjustable mouths were added to the 1386 shoulder plane (by means of a screw-adjustable toe) and to the 2509 (by the insertion of shims between the stock and removable toe); both these models are still available, as the Record 073 and 077 respectively. The 311, or 3-in-1, is also still made; this can be used as a

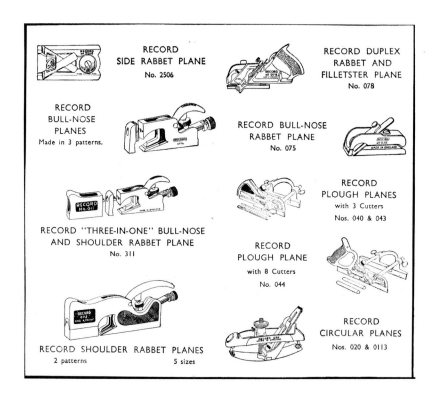

72
Record Planes: group of special purpose planes from a 1939 leaflet. The side-rebate and shoulder planes on the left are all Preston designs.

73
Record Stay-Set and Marples wood planes: the Stay-Set cap-iron is shown dismantled (as if for sharpening the cutter); the plane itself is a standard 07 jointer, apart from the 'SS' logo on the lever cap. The Marples wood smoother was one of the last attempts to make a Bailey-type plane in wood, and dates from 1963.

standard shoulder plane or as a bullnose or chisel plane, depending on whether it is fitted with the long or short front section or none at all.

Record offered their models of the Stanley-Bailey bench planes in all sizes except No. 1, and with the option of a corrugated base (the Record catalogues did not retain the more picturesque American term 'bottom', although they still call try planes 'fore and jointer'). An additional option was the 'Stay-Set' plane; this had a two-piece cap-iron of which the lower half could be lifted off for sharpening and replaced afterwards without the need to re-set the cap-iron as a whole.

The removable piece was located on the upper half by a groove and two pins, and was held tight against the cutter when in the plane by the lever-cap. Planes so equipped were distinguished by an 'SS' logo in the lever-cap. Although introduced in the 1930s and offered for many years after the war, this useful device is no longer made. Its one shortcoming was the ease with which the removable piece could fall into the shavings that normally cover a workshop floor.

Other Record developments from the Stanley range were logical downward movements in the field of ploughs. The Stanley 50 (which Record also made as the 050, with a screw adjustment to the cutter in place of the lever found on later Stanley versions) had two skates because it had originally been designed for beading. Record produced a slightly smaller version with no sliding stock (the cutter was held in

74
Record 044 plough: this example, bought by the author's grandfather in 1939, has never been used.

75
Stanley No. 13 compass plane; an early example, as
indicated by the flat adjustment knob (replaced by
the familiar recessed pattern in 1872–3). The No. 13
only appeared in 1871, and up to 1879 the sole was
attached to the body by screws; later examples
have a keyed slip in the body, riveted to the sole.
(Christie's South Kensington)   (P)

place by a simple screw lever) for plough cutters only; this they called
the 044. Even smaller versions, with no handle as such, were the 040
and 043. These have now gone out of production, while the 044 and
050 have been redesigned with streamlined looks and open plastic
handgrips. These (and later 405s) have circular knurled thumbscrews
in place of the flat type formerly used – a return to the very earliest
Stanley combination planes.

So close was Record's copying of Stanley designs that they even
produced both the Stanley types of circular plane. Of these, the 113
was the earliest design, and it is surprising that Stanley continued to

76
Stanley 113 compass plane: the screw lever cap and
unusual side-wheel cutter adjustment had given
way to the normal Bailey arrangement by 1898. The
113 was introduced in 1879.

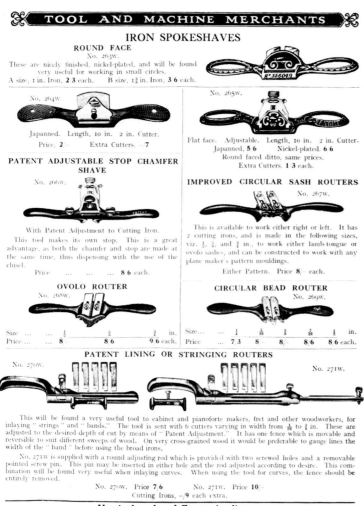

produce it for so long after acquiring the Victor 20, which is regarded as the simpler and more robust design. Admittedly, surviving examples suggest that the 113 was the more popular model in the early days, but neither survives today in Stanley form, and only the 20 (as the 020) under the Record banner.

Record also took up the manufacture of several of the Stanley spokeshave designs. This is interesting in that Preston had long made metal spokeshaves of basically similar type but recognizably different design, to which Record would also have been heirs. The Preston models employed a single, centrally mounted adjustment screw in place of the two screws employed by Stanley; this gave no control

78
Various American iron spokeshaves, some with
wood handles (Melhuish 1925).

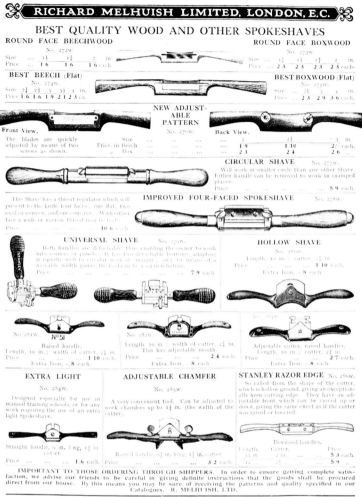

over lateral adjustment, although an ingenious butterfly-lever was
fitted on the more expensive versions to overcome this. It is less easy
to use than the two separate screws, and this may be why Record did
not adopt the Preston spokeshaves, even though their wider, open
handles give a better degree of control to the user. The ornamental
castings would have looked very old-fashioned in the 1930s, but an
undecorated Preston-type stock fitted with the Stanley adjustment
would have offered a spokeshave combining the best of both worlds.
(Preston themselves where already making Stanley-type spokeshaves
by 1930.)

Stanley have recently adopted the two-screw system for a cheap bench-plane aimed at the do-it-yourself market. Other attempts to produce cheaper planes have included the Household and Handyman planes in America and the Acorn range in England. Acorn planes were inherited by Stanley from Chapman's, the Sheffield firm they took over in 1938.

To counteract the tendency for cast-iron planes to crack when dropped on a concrete floor, a number of steel-bodied Bailey planes were offered in the inter-war period, and for carpenters who had to carry their tools around, an aluminium plane would have been a decided advantage. Neither of these types is found very often, and when an aluminium 45 appeared recently in a Christie's South Kensington sale, it realized about ten times the normal price of an iron version. Its rarity was perhaps understandable; the 45 is hardly the sort of tool that would be used outside a joiner's workshop.

4 DECORATED AND DATED TOOLS

The idea of something so essentially functional as a tool being decorated is, in these days, strange and intriguing. Indeed it is so strange that an eminent archaeologist has used the absence of decoration as a criterion to distinguish tools from weapons (D. M. Wilson, *The Archaeology of Anglo-Saxon England*, 1976, p. 255). Even to those who have learned that it was not uncommon, perhaps even customary, for tools to be decorated, the insight into a less strictly functional, less materialist age which the sight of such objects gives exerts a powerful attraction. To express this contrast another way, could we imagine a modern army tank having its turret armour etched with scenes of love and the chase; a machine-gun elegantly shaped as a serpent emerging from a nest of volutes and curlicues; or a guided-missile destroyer with floral gilding embellishing its stern? Just to contemplate such absurdities shows us the distance we have placed between ourselves and our ancestors who did etch their armour, give their guns sculptural form, carve and gild their ships. Yet decorated tools have an extra dimension which contributes to their special interest and appeal. Fine armour, furniture, saddlery and other artifacts were produced for wealthy patrons by specialist craftsmen whose whole *raison d'être* it was to cater for an aristocratic clientele. Tools, on the other hand, were made either by the workmen who needed them for their daily use, or by tool-makers who knew that, for whatever reason, there was a demand that would not be satisfied with purely functional tools produced at the minimum expenditure of materials and labour.

The custom of making tools decorative is clearly an old one as some of the surviving Ancient Egyptian and Roman examples show. Tools of such antiquity are not likely to be met with frequently but collectors should make themselves familiar with the characteristic forms since extraordinary and unexpected finds are the spice of life to those alert enough to spot them. The very rarity of really early specimens makes them less readily recognized for what they are and there can be few collectors without a story of how they found some

79
Roman plane: built up with decoratively shaped iron plates (the empty spaces would have been filled with wood which was hollowed out to provide two hand-holds). The plane is nearly 13 in. long, but only 2 in. wide. It was found in 1880 at Cologne.
(Rheinlandes-Museum, Bonn)

80
Simple crucifix: the cross-bar is given a slant to convey an impression of perspective. It surmounts the fielded throat decoration of a cherry-wood plane which probably originated in Switzerland or Austria.
(John Melville)

ancient treasure labelled as a Victorian kitchen implement and bought it for a song. Examples of such tools from antiquity can be seen in many museums, both national and provincial.

The tradition of decorating, or making decoratively – a distinction which will be discussed later – has never completely died out. However, the two principal motives for doing extra work on a tool, over and above what was strictly necessary to make it functional, seem to have been the need to give it religious significance, and the wish to 'personalize' it in some way. The effect of the first motive, which can be seen on many continental European tools in the form of crucifixes, sacred hearts, or monograms around the letters IHS or INRI, was virtually excluded in Britain by the protestant Reformation. But the urge to give tools some individuality has survived even the dulling effects of mass-production. It can range from simply idiosyncratic name stamps (A. Fake, B. Keen, I am Hastie) to the more elaborate and inventive ideas of which examples are illustrated.

Another influence, quite different from the piety or pride which inspired the majority of 'one-off' designs, made itself felt as soon as tool-making became a specialist industry. That influence was salesmanship. It seems quite clear for example that the famous 'standard' Dutch planes were the product of a few professional tool-making businesses which were turning out their scrolled and whorled patterns at least by the last quarter of the seventeenth century. The number of these planes which are known to have survived from the seventeenth and eighteenth centuries compared with the relatively few undecorated Dutch planes of similar age is convincing evidence that the decorated forms were in normal working use. There may, of course, in addition have been 'de luxe' tools made specially for important and wealthy patrons. Such potential patrons certainly existed, either as enthusiastic amateur craftsmen or as devotees of tools for their own sake. As examples one can cite the sixteenth-century Elector Augustus of Saxony, Count Wrangel in seventeenth-century Sweden and Louis XVI of France, not to mention our own, if humbler, Samuel Pepys. However, signs of wear, replaced parts, and the fact that even the most elaborate examples seldom bear coats of arms or similar indications of an aristocratic commission, suggest that they were first and foremost tools for trade use and that their embellishment was an aid to selling them.

The standard Dutch planes are the best known series of such commercially produced decorative tools. The following planes were included in the range: the *gerfschaaf*, a scrub plane of the distinctive scroll shape which seems to have been popular in many European countries, including England, between the sixteenth and eighteenth centuries; the *roffel*, equivalent to the English jack plane; the *voorlooper*, a fore plane; the *reisschaaf*, a try-plane or jointer; the *blokschaaf*, which seems to have been a smoother but often gets confused with the *bossingschaaf*, with cutter either skew or square, which was a panel-raising plane. There were also the *ploeg* or plough, and a simple rectangular plane about twelve to fifteen inches long,

81
Details of three Dutch standard planes showing different styles of throat decoration and date carving. The 1792 is quite untypical and it shows the sort of variation from the standard pattern which can be encountered. This plane also shows clear evidence of having been modified in order to take a double-iron.
(Science Museum)

without handles, which was probably the Dutch version of the strike block.

With their elaborately shaped handles, carved detail on throats and elsewhere, and relief-carved dates, these planes naturally fetch good prices. Still, considering them as representatives of a whole cultural tradition that have an important place in the development of the modern range of bench planes, in addition to their intrinsic attractions, they seem cheap in relation to the four-figure prices which are paid for rare Stanley models or eccentric versions of the Ultimatum-type brass-frame brace. The rarest of the Dutch standard planes are the *roffel* and the *gerfschaaf*, which is what one would expect since these are the planes which normally get the hardest use. The commonest are the *bossingschaven*, the survival of which parallels the large number of English panel-fielding planes, from Robert Wooding onwards, which are still around while contemporary jacks and smoothers have all but disappeared.

Although the impression gained from seeing a series of these Dutch planes is of a completely standardized product, as alike as castings from a mould, a close side-by-side comparison will show that they are each free-hand interpretations of a basic pattern and, moreover, that some show considerable variation. Particularly notable amongst these variations are the style of the date figures, some examples being in an almost cursive script, and the inclusion in some cases, within the date cartouche, of initials. An example of the latter was the *reisschaaf* sold by Christie's South Kensington on 17 September 1981. It had the initials WVV over the date 1753 and it must have been specially ordered from the maker since the initials were an integral part of the design.

Other series of tools with a conventional and more or less standardized decorative form are the famous Nuremberg etched iron tools of the sixteenth century; an English range of socketted hammer and hatchet heads, forged with intricate spiral decoration and dating from the last half of the seventeenth into the early eighteenth century; planes of cast iron or brass with finely scrolled wooden handles from Scotland; the nineteenth-century Austrian drop-forged axes which bear the traditional patterns of punched stars, flowers and 'sickle' marks; and, ultimately, the arabesqued castings of early Stanley planes or Davis levels from America. Once again it might be thought that such classic museum pieces as the etched Nuremberg tools would be too rare to be of much interest to present-day collectors but at least three examples of this sixteenth-century work have turned up, separately and more or less unnoticed, within the last few years. Others can be seen in museums in Paris (Musée de Cluny), Rouen (Musée le Secq des Tournelles) and Dresden (Kurfürstliche Kunstkammer).

By far the most frequently met decorative tools are 'one-off' jobs. These, one feels, have been made by workmen amusing themselves

82
Ornate hammer: of the type which, from named and dated specimens, we know to have been made in England in the late 17th and early 18th centuries. It is 8½ in. long overall but most of this length is accounted for by the elaborately forged socket. Only the last 2 inches is a turned wood handle. (Christie's South Kensington)   (P)

83
Punched decoration on a south German broad-axe:
two different smiths' stamps can also be seen, each
struck twice. The tool has been painted black and
the decoration picked out in gold. This was
probably done fairly recently for processional use
whereas the axe may well be 16th/17th-century.
(John Melville)

through winter evenings, or during slack periods in the workshop,
giving the tools their own unmistakable imprint and creating a talking
point amongst family or workmates. Such tools are of course harder to
attribute to a particular place or period (unless they happen to bear a
date) than the standard ones, but there are some pointers which can be
looked for.

Wood

The timber used is one such guide. Although it cannot be said that one
species of wood was used exclusively at any particular time or place,
certain species are characteristic of certain areas and identifying them
at least gives a likely starting point. Thus *cormier*, a variety of service-

84
Stanley Miller's Plow: the decorative tradition
revived in the metal combination planes of late
19th-century America.
(Christie's South Kensington)

tree wood, was usual for best quality French planes. Birch was common for Scandinavian tools, but was not uncommon for North American ones. Evergreen oak was a favourite in Italy, Spain and parts of southern France. Hornbeam, latterly often with an addition of lignum vitae to increase the durability of plane soles, was, and still is, quite usual in Germany. Swiss and Austrian tool-makers too used hornbeam but cherry, maple, pear and apple were also popular in those regions. Several planes made from a magnificently dark and heavy wood (possibly cocobolo or some other tree of the genus *Dalbergia*) although found in an old workshop in Switzerland turned out to have been made in Buenos Aires, which is as good an indication as any that tools have been travelling quite far afield even before the recent mass migrations caused by the vagaries of the collector's market! Beech is almost the only wood used for British planes, at least in the last 300 years, and a whiter variety of the same timber was most frequently employed for North American planes. It is usual for one individual plane to be made from a single species of timber. Often, in fact, planes were shaped from a single piece of wood, even when the design, incorporating perhaps a long projecting handle, makes this seem a cumbersome and wasteful procedure. However, it is curious to note that the seventeenth/eighteenth-century standard Dutch planes had walnut handles set into beech bodies – exactly the same combination as the late pattern Norris planes in England. One wonders whether the Dutch planes were originally also heavily stained and varnished in order to conceal the contrasting timber colours.

Type and Shape

After trying to identify the timber we can consider the general type or shape of the tool, but it must be pointed out that no firm conclusion can be drawn from this 'typological' approach unless one is fairly sure of the period to which the tool in question belongs. Although the number of positively identifiable specimens is so small that generalizations are dangerous, the variations in type and general shape of tools from region to region in Europe do not seem to have been so pronounced in the Middle Ages as they became later. For example, most planes in European countries appear to have had front handles until the sixteenth/seventeenth centuries. Then English planes lost this feature (until it was re-introduced from the New World by Record and others copying Stanley!) and the trend was followed in other countries until only the characteristically 'horned' planes of the German-speaking and Scandinavian countries now retain it. So a plane with an upstanding front handle, if made in the last 150 years, will almost certainly be from that area. If it was made in the 150 years before that then France and the Netherlands are also a

85
Boxwood carving chisel handle: carved in 1914 by Mr George Marley (who was then engaged on making wooden saddle trees for the British army). His inspiration for the design came from an illustration of a piece of sculpture entitled 'Dawn', but, as he characteristically commented, he called it 'Morning After'.
(John Melville)

86
Another tool decorated by Mr Marley, in 1965, when he was 85 but still doing a full day's work six days a week. The tool was a brand-new imported German rebate plane with hornbeam sole jointed to beech body. The wavy joint-line inspired the idea of water and the rest illustrates one of Mr Marley's favourite songs, 'The three little fishes and the momma fishy too'.
(John Melville)

possibility. If it is earlier still then it could have originated almost anywhere, including England. This method of attributing tools to areas according to their type is complicated by two further factors. Firstly, some countries did a considerable export trade in tools and they manufactured the types that their customers wanted. Thus, for example, there are Sheffield stirrup adzes and other tools made for the Spanish and South American markets, German tools made to English patterns, and Chinese tools faithfully copying German models. On the other hand, some foreign types of tool enjoy a certain popularity from time to time. Hence the British-made planes, with horn-like front handles, used for roughing and known as 'bismarcks' in recognition of their German inspiration.

In the case of planes the cutting iron may provide some clues. Often the iron will be a replacement, or will be missing altogether. Even so it will be possible to see whether the plane is adapted to use a double-iron and, from the roughness or otherwise of the slot chiselled out to accommodate the bolt, to estimate whether this was done when the plane was first made or subsequently. The earliest documentary evidence for double plane-irons is about 1730 but many single-iron planes were made after that date. Most plane-irons carry a maker's mark of some sort and this, if the iron seems to be original to the plane, will add some evidence. However, here again, it must be borne in mind that the well known English and French edge-tool makers in particular had a wide export trade. Irons bearing inscriptions such as

'Warranted Good Quality', with English-looking crowns and such devices but no maker's name, although found in foreign planes, may have been the product of indigenous manufacturers seeking to exploit the Sheffield reputation. It is worth pointing out here that many of the more primitive, blacksmith-made, plane-irons taper in width from the edge towards the top. This possibility should therefore be considered before any force is used to remove an iron from an old plane. Such irons have to be inserted and removed through the plane's mouth.

87
Old woman's tooth planes: wooden routers are usually home-made and sometimes very rough and ready, but these are beautifully carved. The smaller one is dated 1667. The larger one is Swiss. Both are in the Science Museum.
(Science Museum)

Decorative Styles

When we come to the question of decorative styles we can make a distinction between tools which have been made decoratively and those which have been decorated. In the first category there are tools whose whole form has been designed for its visual effect and also those where the decoration may be more superficial but is an integral part of the original design. In contrast to this group there are tools where the decoration is purely a surface treatment and may well have been carried out by a hand other than that of the original maker. This second category is not necessarily any less deserving of admiration – the judgement must be primarily an aesthetic one – but it does include a growing number of late and less worthy 'improvements' to simple tools which would have been better left as they were. In some instances such operations have been performed with the intention of deceiving as to date or provenance in which case, of course, the result is a fake, to be shunned at any price. It is the intent to deceive which must be the criterion. There would seem to be no reason why today's workmen should not be encouraged to make and beautify their own tools. Indeed, unless some do so, future generations will see even less to admire amongst the products of the twentieth century.

How do we detect the fakes? it can be difficult. In the case of iron (for example, a fine pair of French carpenter's compasses with fleur-de-lys hinge and points springing from dragons' mouths) if the object has been skilfully forged from wrought iron and matured in the

88
Naturalistic carving: this is on the front section of an otherwise ordinary try-plane. The design is possibly a pun on the owner's name.
(John Melville)

compost heap, it is probably impossible. Wood is slightly easier, thanks to the difficulty of imitating the genuine patina of age and wear. But then a fine old wooden tool may have been stripped, or just 'cleaned' in the horrifying ways still recommended by some authors. It may then have been treated with one of the stain-cum-wax preparations guaranteed to give an instant patina. Fine old tools are not so common that one can afford to pass them by just because they look phoney!

Another favoured indicator can be genuine worm-holes which are supposed to emerge more or less at right angles to the surface of the timber. Consequently, holes emerging obliquely or even running along the surface, should be signs that the timber has been re-worked, but it is a remarkable fact that many of the finest old planes have been made from flawed timber. Knots, contrary grain and holes made by a variety of wood-boring insects were clearly features of the piece of wood before it was fashioned into a plane. The most plausible explanation for this is that prime timber would be reserved for customers' jobs and that tools therefore had to be made from offcuts and rejects. Another possible explanation for worm-holes running along the surface has only recently come to light with the realization that some carved tools were originally painted, just as figure carvings often were. The grub, eating its way to the surface, would encounter a layer of unpalatable paint and therefore turn and follow the underside until it reached an unpainted area. So there are no easy answers and each case must be judged on its merits, but one is justified in looking suspiciously at carving which is shallow and rough; at poker work; at apparently primitive art applied to factory-made tools; and at that hard and gritty dressing, favoured by the old school of reproduction makers, which was compounded of glue-size and dust from the workshop floor.

Styles of decoration, and favoured forms for profile or relief, do tend to follow certain conventions even when a tool is a unique, one-off, job. Thus the use of simple punches – dots, stars, crosses, semi-circles and eyebrows or sickle marks – to build up a repeating pattern is characteristic of German-speaking countries but seems unknown in Romance ones. Quite elaborate and attractive effects could be achieved in this way. For example, two rows of depressions, facing in opposite directions, made with a triangular punch, the apex being struck deeper than the base of the triangle, producd a ropework

89
Not a fake: a try plane with just the sort of shallow carving and open worm channels which arouses suspicion of faking. In fact close inspection reveals traces of polychrome painting within the detail and it leaves no doubt that this is a genuine tool of considerable age although its provenance is unknown.
(John Melville)

90
Two south German planes showing typical punchwork decoration allied to shallow carving. One is dated 1774 and the other 1781. (Science Museum)

border. If this ropework were done in the bottom of a shallow groove which had been ploughed out with a coarse-set, and therefore chattering, plough-plane the result was a kind of Florentine braiding. Flowers, foliage and tendrils could be built up using semi-circular punches, or gouge impressions, of various sizes. These flat designs might be supplemented with rows of gouge cuts which give sharper light and shade patterns.

91
Dutch standard planes: *top to bottom, left to right*:
REISSCHAAF, GERFSCHAAF, VOORLOOPER, ROFFEL,
BOSSINGSCHAAF, PLOEG.
(Photo: John Melville)

92
Early mitre plane and etched iron plane: the iron
plane in the foreground is a typical representative
of 16th-century Nuremberg work. The cutter,
marked *Maillot à Lyon*, was probably replaced in
the 18th or 19th century. Two similar planes, from
the Elector Augustus' collection and in unused
condition, toured the USA in 1978 as part of the
Dresden exhibition. The mitre plane, thought to be
English, is clearly a forerunner of the more familiar
English mitre planes; as a type it is thought to date
back possibly to the 16th century.
(Photo: John Melville)

93
Four Dutch bossingschaaven, one with compass
sole. They are dated 1746, 1756, 1766 and 1776.
(*Old Woodworking Tools*)

94
French decorated croze: incised with the Habsburg double eagle with a punched ropework border below. (A croze is used by coopers to form the groove round the inside edge of a barrel into which the lid fits. Most English examples are much plainer, and have a saw-like blade instead of a separate cutter and nicker, as here.)
(John Melville)

South German and Austrian broad-axes have a long tradition of being decorated with punched designs, sometimes floral but often in what appear to be astronomical constellations. These designs together with the smith's trade-mark, which itself is frequently picturesque and may take the form of a tree or an animal, give a rich surface to these massive and rather sombre tools. Older specimens may have a number of different smiths' marks recording their visits to the forge each time they have been re-steeled. The design usually centres on the lower corner of the 'beard' on the upper, bevelled, face of the axe-head (the axes are found right- and left-handed in almost equal proportion) but in some cases there is also a border of fine punchwork along the eye-weld ridge. It should, perhaps, be stated here that three holes, piercing the blade in a triangular formation, may or may not represent the Holy Trinity but they certainly do not prove that the instrument was an executioner's axe. However, tool collectors lucky enough to possess an axe with this feature would be well advised not to publicize this information since many arms-and-armour experts are willing to bid highly at auction for such an object having convinced themselves that the three holes are a sure sign of a sanguinary history! Decorations similar to those commonly applied to broad-axes are also sometimes to be found on froes, picaroons, draw-knives and other iron tools.

Another regional form of decoration was low-relief carving, typically with floral subjects, which was finished with a notably

95
Inlaid plane: dated 1756, this is now in the Science Museum, London.

96
Another view of the 1756 plane, showing how inlay was also used on the sole. There was originally an additional piece in front of the mouth but this is now missing.
(Science Museum)

rounded and smooth surface. This style seems to belong essentially to the Tyrol.

A totally different approach to the decoration of planes involved inlaying pieces of bone and coloured woods so as to make up geometric patterns. Strangely, the inlay work is not confined to the top and sides of the plane but extends over the sole as well. Yet most examples do show the wear which one would expect on a working tool – there is a parallel here with the Nuremberg planes which are

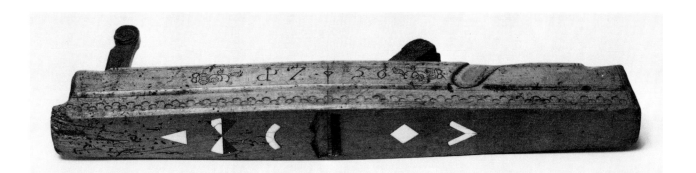

97
Pair of compass planes: dated 1740, and inscribed with the word 'Nigst'; these two planes are thought to come from the Alsace region.
(*Old Woodworking Tools*)

etched on all surfaces. These inlaid tools are often believed to come from Italy but there does not seem to be any firm evidence to support the belief.

Several planes from more northerly parts of Germany incorporate human figures, usually with a slightly comic mien, whereas French workmen in a number of instances have chosen to carve serpents or animals in full relief. Some of the latter, for instance, the snake plane in the Musée des Arts Decoratifs at the Louvre in Paris, verge on the grotesque and one must doubt whether they could ever have functioned well as tools. In the particular case of this plane the shavings have to escape through the snake's mouth and it must have been a problem to clear them if they choked it. Possibly the plane, which is rather small and of very dense wood, was only intended for working end grain.

In the eyes of many collectors the classic forms, when finely detailed and executed, are to be admired above the more exuberant and eccentric pieces, however eye-catching the latter may be. Thus, for example, some planes of the scroll-form, of which the Dutch *gerfschaaven* are the best known exemplars, have such gracefully extended necks and such finely carved and flowing scrolls at the top of the front grip and where the butt of the heel curves under, that it is hard to imagine a man-made object more satisfyingly adapted to its purpose. Another case is the classic French *varlope*, made in a free conformity to the design illustrated by Jacques-André Roubo in 1769. The elements of this design: the curled front grip, the 'flat-iron' rear handle, the bolster backing the cutter, although all carved from a single piece of wood, are shaped as if they were individually joined to the stock. In a good example the carving which conveys this impression of separate yet solid foundation is crisp and elegant. Similar virtues can, of course, be seen in the chamfers and general proportions of English moulding planes before the pattern began to degenerate at the end of the eighteenth century.

Although they are not yet frequently met with in the West a word should be said about oriental tools. There is a growing vogue for importing Japanese tools, for use by those cabinet-makers who appreciate their peculiar qualities, so it is likely that collectors also will extend their scope to include them. To take what is probably the oldest tradition first, Chinese/Korean planes have a gently curving

98
Exotic planes: *left to right*: a plane from the South Seas, carved to represent a timber-chewing beast of that region; plane carved to represent a lady's shoe, probably from France or the Iberian Peninsula; carved plane thought to come from Indonesia.
(*Old Woodworking Tools*)

99
French plane in the form of a snake (Musée des Arts Décoratifs, Paris).

sweep over their upper surface which, after being polished by handling, does show off to best advantage the fine, hard woods from which they are often made. Chinese craftsmen seem recently to have abandoned the old units of measurement with the result that old-style rules can be obtained fairly easily. These have their markings in the forms of little brass studs, set flush into the surface and making an attractive colour contrast with the lustrous sheen of the wood. As there were also Chinese characters on the specimens which the author acquired he went to some lengths to have them read, in the expectation of learning something about makers or dimensions. It turned out that they simply honoured a Confucian tradition by expressing some such sentiment as 'A thousand sons: much happiness'!

Not a lot seems to be known about Indian tools. India is probably one of those countries where craftsmen have not enjoyed the status and independence which are conducive to the production and preservation of fine tools. However, there is an old tradition of brass casting and there is reason to believe that some unusual and decoratively shaped brass planes which turn up in Britain from time to time may have originated in India.

When we come to Japan we have a country where craftsmen have, for many centuries, been held in the highest esteem. They have therefore had the means to acquire fine tools and the pride to preserve them. There has also been an outstanding cutlery industry, if so prosaic a term can be applied to the sublime art of making a Japanese sword blade, to develop the metallurgy and skills needed for the production of edge-tools of the highest quality. Consequently it is first and foremost for their metal parts that some Japanese tools are to be prized so highly. The best chisels and plane irons not only have an

100
The basic form of a 17th/18th-century try-plane (*varlope*). The detailing, e.g. the scrolls and flowers in front of the rear handle, would be individual to each plane.
(John Melville)

101
Traditional Japanese ink-reel: used for marking
straight lines, with a length of inked string
unwound from inside the case. Modern versions
retain the old shape in plastic.
(John Melville)

intricate micro-structure which gives them a glass-hard, dense and
homogeneous cutting edge allied to a fibrous, grainy body, but they
are then filed, hammered and engraved in order to create attractive
effects on their visible surfaces. The wooden bodies of planes are so
low (it puzzles Western fingers simply to grip them) and severely
rectangular that there is no more to admire in them than in any choice
block of wood. One tool which does give its makers scope for the
exercise of imagination is the ink-pot, the Japanese equivalent of the
European reddle-box or chalk bag for striking straight lines. These
may be carved in accordance with various traditional designs (the
hairy-tailed tortoise; the Bhuddhist wheel of life) or plated with
embossed gold, or richly lacquered.

The incidence and style of dating on tools seem to be largely a
matter of cultural tradition. In Alpine regions a village carpenter
employed nowadays to make a cupboard will very likely put his
initials and date on it as a matter of course, and this seems to have been
the habit for hundreds of years. In total contrast, an English
workman, even after spending many hours making a piece of
furniture, will send it out without any indication of date or
authorship. This may not always have been so. The author knows of
nearly a hundred English, or Anglo-American, tools bearing dates
from the seventeenth and eighteenth centuries. There are, as yet,
absolutely no examples available to show us what the situation was in
the preceding period. Nevertheless it is rather rare to find dates on

102
Spelk plane from the Alpine area showing how the figures of the date (1821) were customarily separated into two pairs. (Spelk planes produced long, thin slivers of wood for use as lighting tapers or possibly as 'chips' for weaving into baskets.) (John Melville)

English tools and a large proportion of those that are dated are planes made, or acquired, by various members of the Bottle family, builders in Kent for over 200 years. These dates are composed of individual digits, punched into the end-grain surfaces, and it is evident that the family owned several sets of figure punches over the years. Other English planes, for example the S. Tomkinson moulding plane dated 1779 and illustrated on p. 42 of W. L. Goodman's *British Planemakers from 1700*, or the Darbey/Briscoe plane dated 1778 and sold by Christie's as Lot 160 on 17 September 1981, are dated on the side. It is therefore advisable to look carefully for marks all over a tool, particularly since punched figures do not take clearly on the long grain of plane sides. The surviving seventeenth-century English tools

103
Spelk plane dated 1615: the vertical hand-grip may be a later addition, but the two horizontal grips represent a tradition going back to Roman times. (Science Museum)

are mainly squares and rules. Here again the dates may be small and not immediately visible.

Dates on French and Spanish tools seem even rarer than on English ones, but in the Dutch, Scandinavian and German-speaking countries they often formed a prominent part of the decorative scheme. The Dutch standard planes, with the date carved in relief within the throat decoration, or in a special cartouche on the side of the fence in the case of ploughs, have already been mentioned. Germanic planes usually display their dates conspicuously enough but there are two points worth mentioning in order to help ensure that they are correctly read. First is the fact that the carvers abhorred the simple figure '1'. Consequently, although the capital letter 'I' was often acceptable, plain and unadorned, '1's are usually entwined in their own extended serifs, sometimes to the extent of looking more like 8s or 6s, or even the capital letters J or A. Secondly, the carver usually preferred to arrange the date symmetrically, separating the first pair of digits from the second pair with initials or some decorative motif in between.

The collector who decides to specialize in decorative or dated tools is unlikely to amass the sort of accumulation which tool collectors with more omnivorous tastes sometimes find themselves amassing. Although he may find himself paying a considerable sum for a choice specimen, he has the satisfaction of knowing that he is being discriminating. With care he can be sure that he has an object of genuine historical and technological interest which can be precisely attributed to place and period. Above all, since his choice must be based largely on aesthetic considerations, he is backing his own taste.

5 BRACES AND OTHER DRILLING TOOLS

Of all the tools that have been devised over the centuries for boring holes, there is one pre-eminently associated with the woodworking trades and available in large numbers to the collector, and that is the brace. It seems strange that so simple and obvious a tool should not always have existed, but Goodman records no firm evidence for its use earlier than the fifteenth century.

The majority of woodworking braces made up to the late nineteenth century were of wood; iron braces existed, but were seemingly little used in joinery workshops. As with planes, beech was the favoured wood in the British Isles, but continental braces were made of various hardwoods.

Many early braces (and, in certain specialized trades such as chair-making, late ones as well) were permanently fitted with one size of bit (such braces are often referred to as bitstocks). Finding a means of fitting interchangeable bits securely in a wooden brace was not easy, and before the advent of efficient metal chucks in Great Britain, probably in the eighteenth century, the most common solution was a compromise in which the bit was permanently fixed in a square wood shank or 'pad' which fitted into a large socket in the brace itself. This gave enough grip to transfer the considerable torque of the brace to a relatively thin metal bit; the pad was often prevented from falling out when not in use by a metal or wood spring clip or a retaining peg.

Braces of this type that are found today tend to be of continental origin. There is considerable variety of design, with braces from the German and Scandinavian countries being generally rather heavy and angular in outline, with thick webs (where the greatest strain occurs across a section of short grain) often ornamented with mouldings or even curved, in the manner of eighteenth- and early nineteenth-century crank-handles (to which the brace is, of course, related).

In contrast to the chunkiness of some of these braces is the French chairmaker's brace, which is of much lighter construction and curved gracefully to a point at the top, which is designed to fit in a conical hole in a wooden pad strapped to the user's chest. This was one

104
Variety in braces: *top row* (*left to right*). Continental
pad brace; South German pad brace with revolving
handle; English plated button-chuck brace; Dutch
pad brace; *bottom row*, iron brace with elm head;
Henry Pasley's 'Ne Plus Ultra' brass-framed brace;
Scotch iron brace; bow drill using the same stock as
a bevel-gear drill (*all from the Arnold & Walker sale,*
1979).
(John Melville)

efficient solution to another long-standing problem for the brace-
maker: how to provide a durable bearing at the head while at the same
time reducing friction to a minimum. A similar shape is also found
with a built-in head, and slightly flatter webs, in Holland. The
asymmetric shape, with the upper web sloping steeply up to the head,
makes for greater strength in that part, while keeping the hand-grip as

105
Early button-chuck brace: distinguished from later examples by its iron chuck, with typical fine moulding around the nose. Otherwise, the brace is similar to the basic English type current throughout the 19th century.
(*Old Woodworking Tools*)

close as possible to the 'business end', thus minimizing any loss of power through the natural flexibility of the wood.

Dating of these 'primitive' wood braces is not easy, as little positive information exists, but the great quantities surviving on the Continent suggest that many are of nineteenth-century date, although made to a traditional pattern. In Great Britain, the tradition survived to some extent in specialized trades such as coopering and chair-making, where braces were apparently usually home-made, and had permanently fitted, or at least not easily removable bits, since there was little variation in the sizes of holes drilled. However, the mainstream of development in English and Scottish braces in the nineteenth century lay in the metal-reinforced wood brace with brass chuck. This appears to have been established by the end of the previous century, with wide chamfers to the webs and iron chucks containing a spring-loaded catch released by pressing a button. The main part of these chucks is hexagonal to match the chamfered wooden stock, but has a coved circular nose which, in early examples, often has a series of narrow mouldings in the tradition of cold-worked metal of that time. Forging this chuck from iron and then filing it to

106
Braces in wood and iron: at the top is a very unusual, large wood brace thought to be English, 17th-century, although very little is known about it. It is 28 in. long. The three projecting arms could be used for extra leverage, or would give a centrifugal effect if the brace were being turned rapidly in the normal manner. In the centre is a typical Dutch pad brace, and at the bottom a 'sixpenny', 'common' or 'ball' brace, a type common in British tool catalogues throughout the 19th century. It is smaller than usual, at only 8½ in. long.
(*Old Woodworking Tools*)

provide the final, close-fitting shape must have been an expensive business, and from about the beginning of the nineteenth century, brass carvings were used instead, with iron-lined sockets.

One of the earliest known makers of these chucks was Benjamin Freeth, a Birmingham tool-maker of the 1770–1824 period; his name has been seen on iron and brass chucks, an interesting fact in itself since the practice of stamping makers' names on chucks did not apparently become widespread until nearer the middle of the nineteenth century. When names do appear on early braces of this type, they tend to be stamped into the end-grain of the upper web, and are usually those of well known plane-makers (including Mutter and John Green). It is likely that these braces were bought in by such 'makers', who were often tool-dealers as well as plane-makers.

Many early examples of button-chuck braces have the button on the side that is uppermost when the brace is held with the handle to the left and the chuck end furthest from the body. This is the position in which the brace would be held in the right hand, while the bit was being changed with the left. The majority of English nineteenth-century braces have the button on the other side, for holding in the left hand, but the earlier position was retained by many Scottish makers.

The weak point of all wooden braces was the short-grained section forming the webs, across which the power was transmitted from the user's hand to the bit, and this part inevitably tended to split under the strain. To counteract this tendency, it became customary for the better wood braces to be reinforced with brass plates on the faces of the webs. This practice is thought to have begun at the end of the 1820s, probably with the firm of Brown and Flather of Sheffield, whose name is one of the best known on beechwood braces. (By the time of the Great Exhibition in 1851, Henry Brown and Sons and David Flather were in business separately.) The plates were normally inlaid in the wood and secured with iron (not brass) screws.

The head on many beechwood braces was made of a more exotic timber such as lignum vitae or ebony; often the neck (the stem under the head) was of the same wood, but by the 1840s it was also often made in brass. Ebony was the most frequently used wood for the head when the neck was brass, with a brass 'plug' in the centre to conceal the bearing and retaining nut. This plug could be smooth or embossed with a trade-mark or coat of arms. For the collector specializing in braces, the various forms of swivel joint in the head make a subject for study in their own right. Considerable pressure is exerted on the head when a hole is being bored, and considerable tension when the bit is withdrawn from the hole. In most wood braces, the end-thrust during use is taken by the wide surface between the neck and a circular metal plate on top of the brace's stock, often with a steel washer in between the two. Much less friction would have resulted from a pointed centre bearing, but few makers adopted this; the notable exception was

107
British braces: top, a cooper's brace or bitstock (identifiable by its wide head); bottom, a Scottish button chuck brace with right-handed button: the name Stewart, an Edinburgh planemaker of the first half of the nineteenth century, is stamped on the upper web, just below the neck.
(Christie's South Kensington)   (P)

Brown and Flather again, who employed Henry Brown's patent anti-friction head from the mid-1840s (when the basic patent was acquired from its inventor, one John Bottom, whose name appears as 'inventor' on some of the later Henry Brown and Sons braces).

To retain the head on its spindle when the bit is withdrawn from the hole, a collar was screwed on to the spindle, in a widened recess in the upper part of the head, which was then closed by screwing in the 'plug' referred to earlier. Although a single collar nut was used in basic models, double nuts with right- and left-hand threads were employed in more sophisticated braces. A more ingenious form of collar was that of Fenton and Marsden's 1847 registered design, in which a split steel collar was inserted through a slot in the side of the neck to engage a groove in the spindle; the slot was then closed by a flanged brass ring held in place on the neck by the wooden head, which screwed down on top of it. This neck also incorporated an improved form of centre-point bearing.

108
English braces: *top row, left to right*, brass-framed brace by Colquhoun and Cadman, with ebony infill and lever chuck; plated lever-chuck beech brace by James Howarth, Sheffield; similar brace by the same maker, in ebony; *bottom row*, button chuck beech brace with stamp of Holtzapffel and Co.; brass-framed lever chuck ebony brace by Howarth; boxwood Ultimatum brace with ebony head (stamped 'Ultimatum' but without the normal 'Marples' logo).
(Christie's South Kensington)   (P)

Although the button chuck was very satisfactory and continued in use until the demise of the wood brace early in the twentieth century, in the hey-day of English brace manufacture (the period from around 1840 to 1870) several attempts were made to improve it. The two best known variations are the lever chuck and the ring chuck, both of which consist only of different methods of releasing the spring catch which retains the bit. Possibly the earliest lever chuck was that of Bloomer and Phillips in 1847. In the button chuck mechanism, the spring is *pushed* away from the bit by pressure on the button and therefore has to be on the opposite side of the bit-socket from the button. The Bloomer and Phillips lever was mounted on the same side as the spring, which it *pulled* away from the bit. A revised version appeared in 1851, in which the catch engaging the nick in the bit-shank was formed on the end of the lever itself instead of on the spring, and the spring simply pressed against the rear end of the lever. The extra leverage makes a lever chuck easier to operate than a button chuck, but the difference is nice rather than startling. A similar effect was achieved by George Schofield in 1848 by simply adding an externally pivoted lever to a button chuck, which increased the pressure on the button.

It was in or about 1851 that William Marples acquired John Cartwright's 1849 patent for a metal-framed brace incorporating what is now generally referred to as a 'ring chuck'. This neat-looking device consists of a ring at the nose of the chuck, engaging the end of the spring. The ring has a rectangular hole, so that thumb pressure on the edge moves it sideways even when the bit is in position, thus depressing the spring. The ring was retained on the nose by a nut, but this form was used for a short period and is seldom seen. The version normally found is simpler, with the ring attached directly to the spring, and was the subject of a previous registered design by the rival Robert Marples in 1848. Ring chucks are occasionally found on wood braces, but are associated in particular with the metallic frame braces stemming from Cartwright's 1849 patent.

The best known of these is the William Marples 'Ultimatum Framed Brace'. Whereas brass plates had previously been used as reinforcement for wood-framed braces, the principle was now reversed, and a brass and steel frame formed the body of the brace, with wood used for the head and the handle, and as an infill for the webs. The chuck and lower web plates were a single casting, connected by a steel spindle passing through the handle to the upper web casting. The handle itself was free to revolve, a considerable improvement on most other wood braces where the handle has to rotate within the user's hand.

Ultimatum braces were obviously intended for the 'top drawer' of the woodworking trades: they were beautifully made, heavy (and therefore unsuitable for on-site work outside the workshop) and finished with an inlaid ivory ring in the head. The wood was ebony in,

109
Brass-framed braces: *top to bottom*: ebony Ultimatum, beech Ultimatum and a James Howarth brace with boxwood infill and lever chuck. Both Ultimatums are by William Marples and have ring chucks; the ebony version is in exceptional condition, with the original lacquer on the brass. (*Old Woodworking Tools*)

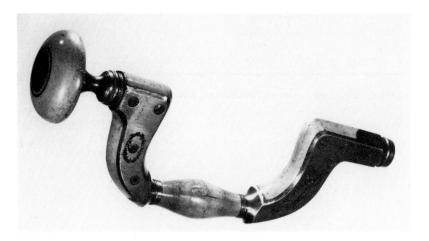

110
Boxwood Ultimatum: this view of the rare
boxwood-filled William Marples Ultimatum brace
shows its distinctive head, with ebony ring.
(Christie's South Kensington)   (P)

perhaps, ninety per cent of examples; rosewood and boxwood were
alternatives at the same price, and beech was also available. The
Cartwright patent mentions the possible use of horn or gutta percha,
and one or two horn-filled Ultimatums have been reported.

In plated braces, beech is by far the most common wood used,
followed by ebony, box and rosewood, probably in that order. In
Ultimatum and other framed braces, the order is ebony, beech,
rosewood and boxwood. The last two may be in the wrong order, but
both are rare; the boxwood models are the more desirable if only
because they are more distinctively different from ebony than is
rosewood. Heads are nearly always in ebony, even on many boxwood
braces, but the Marples Ultimatum in boxwood usually has a
boxwood head to match, with an ebony ring in place of the usual
ivory.

The brass-framed brace was eventually (in the 1860s and 1870s)
sold by many other firms, the majority of them being well known as
makers of wood braces or edge tools. The name Ultimatum was used
by Robert Marples as well as William, but most of the others simply
put their name and a more or less standard legend such as 'Metallic
Framed Brace' on the brass frame. Most of these braces were probably
made by Marples or one or two other firms – certainly, few of the
names appearing on them are those of the actual manufacturers. One
who claimed that his braces were of his 'own manufacture' was Henry
Pasley, whose stamp was a clear imitation of the William Marples
Ultimatum. His brand-name was 'Ne Plus Ultra', which in effect
means exactly the same as 'Ultimatum'. Not all these 'trade' braces had
ring chucks; lever or even button release mechanisms are found, and
not all have an ivory ring in the head.

One framed brace which was significantly different had thick,
central brass webs flanked on each side by ebony, instead of the other
way about. Saw-handle screws and nuts were used to fasten the
sections of wood together. Examples have been seen stamped Sims

and also Stevenson, Mawhood and Co., but in numbers too small for any conclusions to be drawn as to the actual manufacturer. Ken Roberts, in *Some 19th-Century English Woodworking Tools* (which contains the most detailed study yet of the English brace) suggests that Thomas Pilkington was the actual maker, since some at least of these braces have chucks resembling those of the Pilkington braces.

The latter were a strange compromise between the plated and the framed brace; they were essentially of the first type, but with exceptionally large brass plates screwed on to the surface rather than inlaid. The ornamented outline of these plates, typically mid-Victorian in style, makes them very distinctive and highly collectable today, particularly as examples are few and far between. The Pilkington patent was granted in 1852, and the braces were probably made over the next ten years or so. The heads are normally made wholly or partly of brass; in the latter case, the neck expands into a wide flange covering almost the entire underside of the wood cap. Similar heads are occasionally seen on other plated braces.

Another compromise design was Robert Marples' Double Bound Octagon Lever Pad brace, of 1857. (The chuck was still referred to as a 'pad', but I have adopted the modern word 'chuck' in this chapter to avoid confusion with the earlier form of 'pad' brace described at the beginning.) This also had thick, non-inlaid plates, but they were unornamented and looked at first sight like those of the framed models. What distinguished them structurally was that they continued round the ends of the webs, so that the two plates on each web formed a single casting. The only distinctive part of the Octagon

111
Framed braces: details of two makers' stamps. On the left is a standard William Marples Ultimatum, on the right a lever-chuck Improved Suffolk Brace by Thomas Turner and Co.

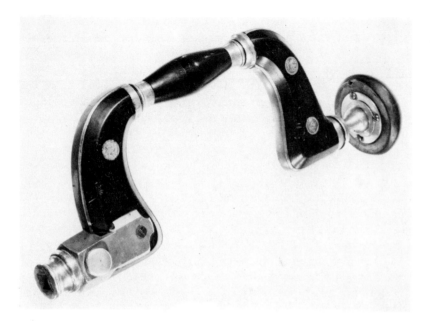

112
Brass-framed ebony brace by Stevenson, Mawhood and Co.: this type of framed brace is much rarer than the Ultimatum type with the wood sandwiched between the brass plates. (The head is a replacement.)
(Christie's South Kensington)   (P)

Lever Pad apart from its octagonal form, was the shape of the lever, a tapered shape working in a slot in the chuck; its design, translated into plastic, would not look out of place in a 1970s electric light switch.

Brass-framed braces continued to be available until the 1890s, and plated and plain wood braces into the first decade or so of the present century. From the 1870s, however, imported American braces forced most of the traditional types off the market. The American braces had

113
Pilkington brass-plated brace: marked 'Pilkington, Pedigorr and Co', 'made for Stanley, Morton and Davis', 'Brass mounted brace'. The design of the brass fittings is typical of the 1850s and 60s.
(W. S. Rhea collection)

efficient chucks which would accept any square-shanked bit (the English button, spring and lever chucks all required bits with notches cut in the right place, which was not standardized). They were also stronger and cheaper. The extra strength was important at the time because much better bits were coming into use, capable of drilling larger and deeper holes than had previously been possible – so much so that the auger became an increasingly uncommon sight in a woodworker's tool kit.

Of the American chucks, the two most successful were the Spofford, in which a socket formed in the end of the brace had a split side with a clamping screw, and the Barber, in which a pair of alligator jaws are drawn together by a threaded outer ring. This type is still in use. Both the Spofford and the Barber have the advantage over most previous chucks that they are self-centring, regardless of the size of the bit shank. Previously, bits themselves had been either of the spoon type or centre-bits. The former were shaped very much like long, thin gouges, and variations included shell bits and nose bits –

114

Twist bits: these two pages from the 1925 Melhuish catalogue show the wide range of twist bits then available singly and in sets, and also some more traditional bits in the 'Technical Outfit'.

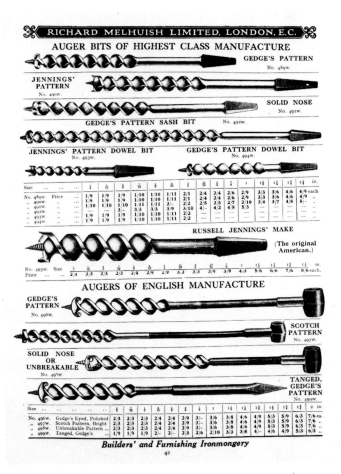

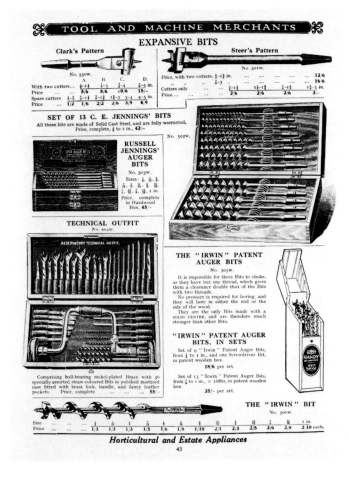

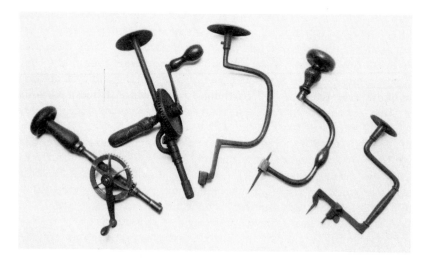

19th-century ironware (*left to right*): bevel-gear drill (wheelbrace) with wrought-iron stem and cast-bronze wheel; breast drill, with wrought-iron stem and head, cast-iron wheel; gas-fitter's brace; 'sixpenny' or plain brace; washer-cutting brace (used for cutting leather washers for well-pumps, for example).

the latter had a right-angled lip across part of the crescent-shaped nose, which was useful in cutting into end-grain. (Because this type, unlike the modern screw-fed spiral bits, does not tend to follow the grain, it remains preferable for boring long holes with the grain – the hole for the flex in a standard lamp, for example.) The centre bit was more suitable for drilling holes of moderate depth in a precise position.

None of these bits was self-feeding, and constant pressure was needed to make them bite into the wood, with frequent withdrawals to clear chippings from the hole. The spiral auger, which automatically cleared the waste as it revolved, was invented in England in 1770 by one Phineas Cooke, but most subsequent improvements occurred in the United States. The most important was Russell Jennings' patent of 1855, for a double-spiral auger bit with a spur on the edge of each thread base. This is the brace bit in most common use to this day; in its larger sizes it would have imposed a considerable strain on an English wood brace.

The use of iron for braces was not new by any means, and the 'Scotch' brace offered in many British catalogues in the nineteenth century was an elegant and well-made joiner's tool, but neither it nor its larger (and less elegant) cousin the wagon-builder's brace could match the American model for price, even though they lacked the luxury of a revolving wood handle and, with the exception of some late examples, a Barber chuck. (The Scotch brace had an elegant, built-in version of the lever chuck, while the wagon brace, like most early iron types, usually had a simple thumbscrew.) Other British iron braces included the 'common' (also known as 'ball' braces from the near-spherical wood handles sometimes fitted, and as 'sixpenny' braces from their cheap price, though it is uncertain when, if ever, they were as cheap as 6d.) and the gas-fitter's – an iron-headed version of charmingly primitive appearance, suggesting an origin dating from many years before there was any gas to fit.

A feature of many, if not most, American braces was a ratchet device enabling the tool to be used in confined spaces, where the crank could not be fully rotated, without reversing the bit between strokes. For working in particularly awkward spaces, various contraptions were devised, including a brace with one crank only, giving a much reduced height, and a universally jointed gadget for connecting an ordinary brace to the bit at an angle. An angle brace, with the brace mounted in a frame at an angle to the bit, was also made, as was a brace incorporating a hand-wheel and bevel gear. Such devices were mainly of use to electricians, plumbers and others needing to drill holes outside the ideal conditions of a workshop, but the electric drill has to a large extent replaced them.

The great advantage of the brace over other boring tools was its continuous rotary motion; the only other hand tool to provide this was the bevel-gear drill or wheel-brace. This is naturally thought of as a modern tool, but it existed in the eighteenth century, beautifully made in the manner of armourer's or locksmith's work. American bevel-gear drills (including the larger breast drills) are made of cast

116
Traditional iron braces and bits: from an early-1920s catalogue of S. Tyzack and Son; the Scotch and wagon builder's braces are updated with ball-bearing heads and even a Barber chuck in one case.

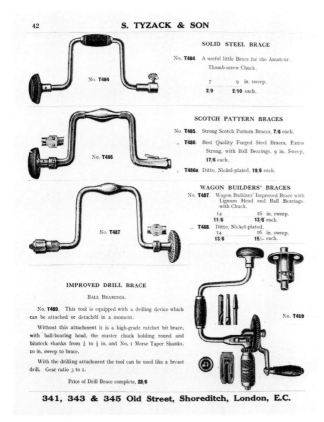

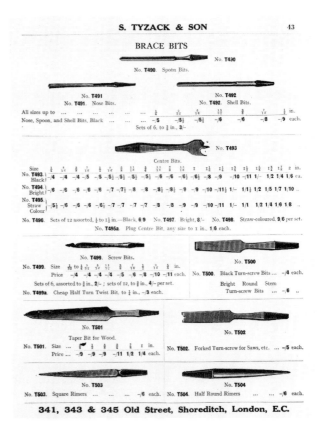

117
Early iron braces: that in the centre bears the date 1735, and the brace above is thought to be late 17th-century, French; the beautiful example at the bottom is probably late 16th-century.
(*Old Woodworking Tools*)

118
Continental drills: the beautifully made iron bevel-gear drill on the left is stamped I. Peter, Lenz, but was probably imported to the U.K. from Switzerland when new in the early 19th century. The brace is of walnut, probably 18th-century.
(Christie's South Kensington)   (P)

119
Millers Falls small wheel-brace: a tiny drill (it is only 8 inches long), and a typical American mass-produced product in cast iron (cf. the English wrought-iron drills on p. 120. The simple chuck will only accept the diamond-point drills supplied with it (so called from their shape) and its ineffectiveness perhaps accounts for its little-used condition.

iron, but wrought-iron versions were made in the late nineteenth century, in the style of the gas-fitter's braces mentioned earlier.

A tool with a very long history was the bow-drill. This was one of the many reciprocating designs, operated by a length of twine or gut stretched between the two ends of a bow and wound round the circular stock of the drill. In the eighteenth and nineteenth centuries, a more sophisticated type of bow was available, consisting of a long sword-like steel blade with a ratchet device for tightening the string. Bow drills, made in ebony or rosewood and brass, are attractive tools to collect although they seldom come complete with bow. Amazingly, they were still in use in the 1920s for the very laborious task of drilling the wrest-pin holes in pianos.

Related to the bow-drill is the pump-drill, which has a long stem passing through a cross-stick attached at each end to a piece of string which is also attached to the top of the main stem. With the cross-stick at the top of the stem and the string wound tightly round the latter, the stick is pressed down so that the unwinding string causes the stem to revolve. As the stick reaches the bottom of the stem and the string is fully unwound, pressure is released, and the rotating of the stem causes the string to re-wind, drawing the stick up again. This all happens much more quickly than it can be described, and a skilled user can keep the drill working with a steady, rapid pumping action.

A more mechanical form of this idea is the nineteenth-century Archimedean drill, in which a sliding collar is moved up and down a spirally fluted shaft. Two variations are the centrifugal drill (which

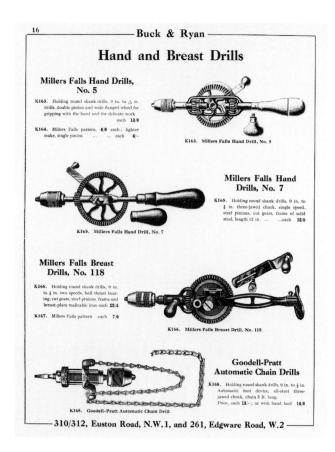

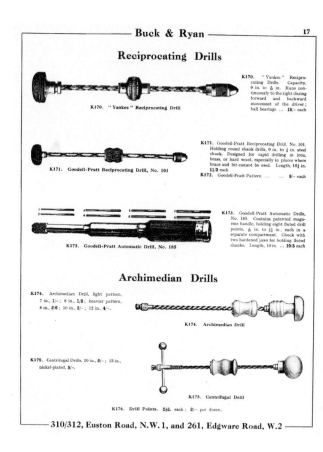

has, just above the chuck, an arm with ball-weights at each end and a ratchet device inside the collar, so that the drill continues to revolve during the idle upward stroke) and the American double-spiral or 'Yankee' type, in which a forward motion is imparted more positively during both the up and down strokes.

120
American bevel gear drills, 1930: the chain drill was used for drilling into pipes, stanchions and the like, the chain being wrapped round the object being drilled and kept in tension by the screw device on the shank.

121
Spiral action drills (Buck and Ryan, 1930).

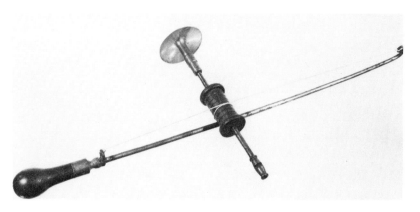

122
Bow drill: unlike many bow drills (and they are by no means common), this one is complete with its bow.
(Christie's South Kensington)

(P)

123
Turning plane and auger: the adjustable turning
plane (also known by various names including
'rounder', 'witchet' and 'stail engine') has a brass-
plated mouth; this one is unusual in having a
maker's name (Greenslade, Bristol). The auger is of
the traditional 'shell' pattern.
(Christie's South Kensington)   (P)

Augers were normally used for boring large holes until the advent
of the American braces late in the last century, and continued to be
used where their greater torque and longer reach were needed.
Although numerous types of auger exist, particularly in the various
patented twist versions, most of these are duplicated in brace bits,
which are smaller and much easier to collect. Augers tend to attract
very limited interest when they appear at auction. Related to the
auger is the gimlet, effectively a miniature auger, which is likewise
hardly used today. The wheel-brace and the electric drill have largely
replaced it, although it still has a place as a household tool where very
occasional need to drill holes does not warrant the expense of
anything more elaborate. Most gimlets have a screw point and a
straight or spiral shell-type shank; English ones usually have a wood
T-handle, but on the continent the shank itself is often bent round at
the top to form a bow. An interesting variation found occasionally is
the bell-hanger's gimlet – a much longer type, up to thirty inches or so
in length, and usually fitted with a slightly oval boxwood handle.

124
Centre-bit: this highly ornamental bit is assumed to
have been made for exhibition purposes. The floral
border is gilt, and the hole it would drill would be
$6\frac{1}{4}$ in. diameter.
(Christie's South Kensington)   (P)

6   OTHER TOOLS

It is no accident that planes occupy so large a part of this book, for there is no doubt that their numbers and the available information on them make them very collectable. However, other tools are just as worthy of being collected, and the formation of a specialized collection is often the first step towards original research in a subject.

The largest remaining group of tools after planes and braces is that of edge tools. This includes chisels, axes, adzes, draw-knives and related implements. Chisels are the most familiar, being still in widespread use in all forms of woodwork from rough carpentry to fine cabinet-making. Sadly, early chisels have a low survival rate, since there is not much left of a chisel when once the blade has worn down to the stub. As with so many tools, it is often the more specialized varieties that do survive, especially heavy versions such as those used by shipwrights and wheelwrights. The typical western chisel of today, with its parallel blade, seems to go back only to the eighteenth century; before that, blades tended to be tapered, so that as they wore down with constant sharpening, the cutting edge became progressively narrower.

Among the heavy and very traditional-looking chisels that do survive is the wheelwright's 'bruzz', a V-form gouge used for chopping out mortices in wheel hubs (which often have acute-angled corners, which an ordinary mortice chisel would not be able to cut cleanly). Since twentieth-century examples look very little different from their predecessors, such tools are always worth examining carefully for makers' marks. Although little has been written about chisels and gouges, their makers usually produced plane irons as well, and information on British plane-iron makers is readily available from the check-list in Goodman's *British Planemakers*.

Another very specialized chisel that survives in large numbers because of its limited use and sturdy build is the lock-mortice chisel. This was an essential feature of every house carpenter's toolkit in the days when mortice locks were six inches or more in length, and were housed in the lock-rail of the normal eighteenth- and nineteenth-

125
Miscellaneous tools: *top to bottom, left to right*: try-square, spirit level, small try-square, bow saw, padsaw, spokeshave, screw-stem plough, sliding bevel, draw-knife, mortice gauge, 'toy' saw, plumb bob, panel saw, hammer.
(*Old Woodworking Tools*)

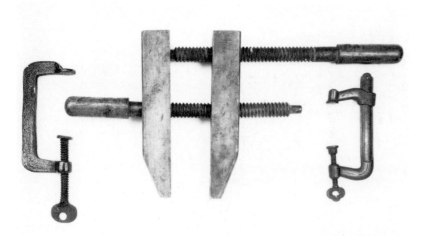

126
Cramps: a wooden handscrew, flanked by a simple wrought-iron G-cramp on the left and a Weston's Patent cast-iron quick-action version by Timmins, Birmingham, on the right.

century panelled door. Cutting away the waste wood at the bottom of a mortice six inches deep and only three or four inches wide is not easy at the best of times, but when the bottom inch or two is in end-grain, a chisel with a bend in it is essential to cut across the grain. Mortice-lock chisels of the 'London' pattern (lighter and without the 'knee' at the base of the curve found on some heavier models) were still stocked by the better tool-shops in the early 1960s, but even then the traditional mortice lock, with the handle behind the key-hole rather than above it, was seldom used except in replacement work.

Although chisels are still being made, the variety of standard types has been reduced, and the firmer and bevel-edged are the two main chisels remaining, often with plastic handles. Although most twentieth-century chisels were sold with handles, the traditional

127
Screw boxes and taps: French and English patterns, for cutting threads in wood.
(Tyzack *circa* 1920)

128
Chisels, from the 1925 Melhuish catalogue.

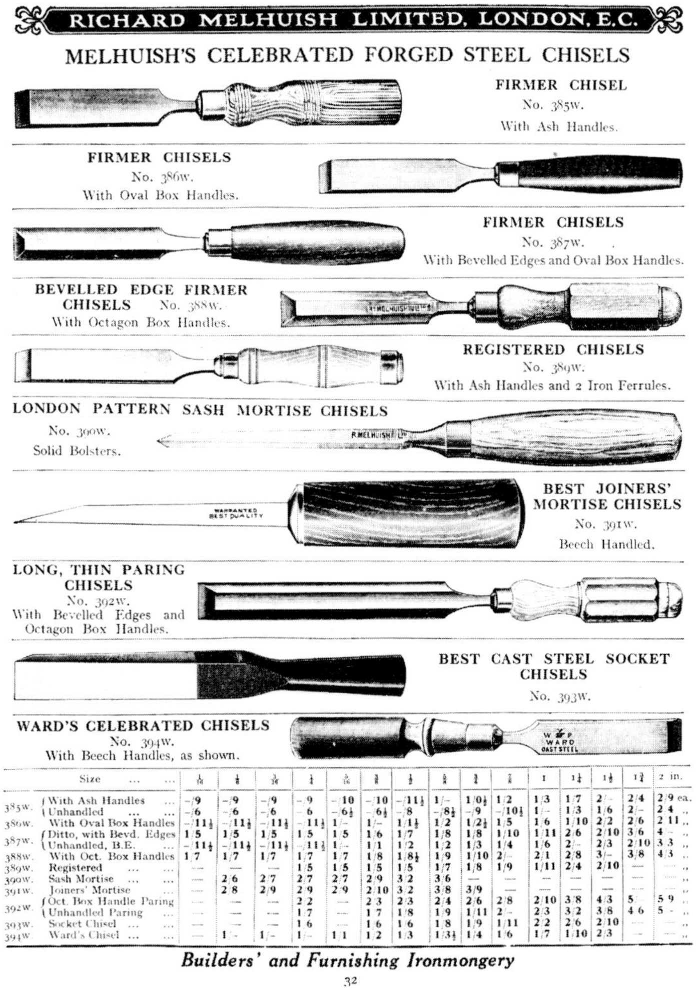

RICHARD MELHUISH LIMITED, LONDON, E.C.

MELHUISH'S CELEBRATED FORGED STEEL CHISELS

FIRMER CHISEL
No. 385w.
With Ash Handles.

FIRMER CHISELS
No. 386w.
With Oval Box Handles.

FIRMER CHISELS
No. 387w.
With Bevelled Edges and Oval Box Handles.

BEVELLED EDGE FIRMER CHISELS No. 388w.
With Octagon Box Handles.

REGISTERED CHISELS
No. 389w.
With Ash Handles and 2 Iron Ferrules.

LONDON PATTERN SASH MORTISE CHISELS
No. 390w.
Solid Bolsters.

BEST JOINERS' MORTISE CHISELS
No. 391w.
Beech Handled.

LONG, THIN PARING CHISELS
No. 392w.
With Bevelled Edges and Octagon Box Handles.

BEST CAST STEEL SOCKET CHISELS
No. 393w.

WARD'S CELEBRATED CHISELS
No. 394w.
With Beech Handles, as shown.

R. MELHUISH Ltd
WARRANTED BEST QUALITY
W P WARD CAST STEEL

	Size	$\frac{1}{16}$	$\frac{1}{8}$	$\frac{3}{16}$	$\frac{1}{4}$	$\frac{5}{16}$	$\frac{3}{8}$	$\frac{1}{2}$	$\frac{5}{8}$	$\frac{3}{4}$	$\frac{7}{8}$	1	1¼	1½	1¾	2 in.
385w.	With Ash Handles ...	-/9	-/9	-/9	-/9	-/10	-/10	-/11½	1/-	1/0½	1/2	1/3	1/7	2/-	2/4	2/9 ea.
	Unhandled	-/6	-/6	-/6	-/6	-/6½	-/6½	-/8	-/8½	-/9	-/10½	1/-	1/3	1/6	2/-	2/4 ,,
386w.	With Oval Box Handles	-/11½	-/11½	-/11½	-/11½	1/-	1/-	1/1½	1/2	1/2½	1/5	1/6	1/10	2/2	2/6	2/11 ,,
387w.	Ditto, with Bevd. Edges	1/5	1/5	1/5	1/5	1/5	1/6	1/7	1/8	1/8	1/10	1/11	2/6	2/10	3/6	4/- ,,
	Unhandled, B.E. ...	-/11½	-/11½	-/11½	-/11½	1/-	1/1	1/2	1/2	1/3	1/4	1/6	2/-	2/3	2/10	3/3 ,,
388w.	With Oct. Box Handles	1/7	1/7	1/7	1/7	1/7	1/8	1/8½	1/9	1/10	2/-	2/1	2/8	3/-	3/8	4/3 ,,
389w.	Registered	—	—	—	1/5	1/5	1/5	1/5	1/7	1/8	1/9	1/11	2/4	2/10	—	— ,,
390w.	Sash Mortise	—	2/6	2/7	2/7	2/7	2/9	3/2	3/6	—	—	—	—	—	—	— ,,
391w.	Joiners' Mortise ...	—	2/8	2/9	2/9	2/9	2/10	3/2	3/8	3/9	—	—	—	—	—	— ,,
392w.	Oct. Box Handle Paring	—	—	—	2/2	—	2/3	2/3	2/4	2/6	2/8	2/10	3/8	4/3	5/-	5/9 ,,
	Unhandled Paring ...	—	—	—	1/7	—	1/7	1/8	1/9	1/11	2/-	2/3	3/2	3/8	4/6	5/- ,,
393w.	Socket Chisel	—	—	—	1/6	—	1/6	1/6	1/8	1/9	1/11	2/2	2/6	2/10	—	— ,,
394w.	Ward's Chisel	—	1/-	1/-	1/-	1/1	1/2	1/3	1/3½	1/4	1/6	1/7	1/10	2/3	—	— ,,

Builders' and Furnishing Ironmongery

32

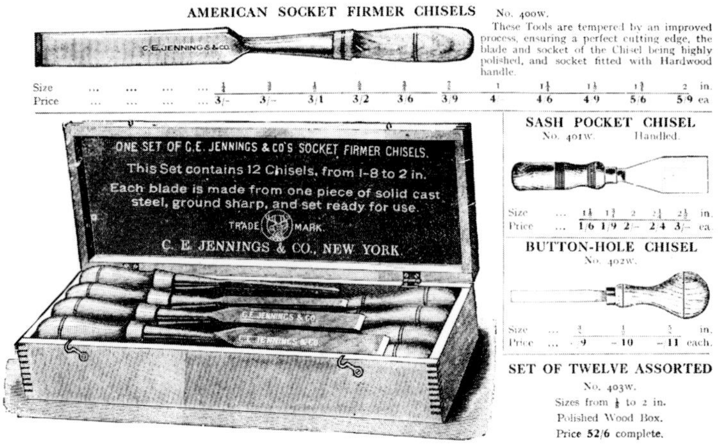

AMERICAN SOCKET FIRMER CHISELS No. 400w.

These Tools are tempered by an improved process, ensuring a perfect cutting edge, the blade and socket of the Chisel being highly polished, and socket fitted with Hardwood handle.

Size	$\frac{1}{4}$	$\frac{3}{8}$	$\frac{1}{2}$	$\frac{5}{8}$	$\frac{3}{4}$	$\frac{7}{8}$	1	1¼	1½	1¾	2 in.
Price	3/-	3/-	3/1	3/2	3/6	3/9	4/-	4/6	4/9	5/6	5/9 ea.

C. E. JENNINGS & CO.
ONE SET OF C.E. JENNINGS & CO'S SOCKET FIRMER CHISELS.
This Set contains 12 Chisels, from 1-8 to 2 in.
Each blade is made from one piece of solid cast steel, ground sharp, and set ready for use.
TRADE MARK.
C. E. JENNINGS & CO., NEW YORK.

SASH POCKET CHISEL
No. 401w. Handled.

Size ...	1¼	1¾	2	2¼	2½ in.
Price ...	1/6	1/9	2/-	2/4	3/- ea.

BUTTON-HOLE CHISEL
No. 402w.

Size ...	$\frac{3}{8}$	$\frac{1}{2}$	$\frac{5}{8}$ in.
Price ...	-/9	-/10	-/11 each.

SET OF TWELVE ASSORTED
No. 403w.
Sizes from $\frac{1}{8}$ to 2 in.
Polished Wood Box.
Price 52/6 complete.

129
Chisels: 18th- and 19th-century examples, including a heavy gouge by Weldon (Sheffield, 1774–88).

131
(*Right*) Wheelwright's tools: between the wheel spokes are two heavy socket-handled chisels, a rounder, a bench-knife and a traveller (for measuring the circumference of a wheel). Outside are two more rounders, a hub auger, a pair of callipers and a socketed gouge.
(Christie's South Kensington)

130
Lock mortice chisels: a selection from the Arnold and Walker sale, including an unusual one with a double neck – presumably intended to provide two fulcrum points, for use in different depths of mortice. The chisel second from bottom is the pattern seen most often.
(Christie's South Kensington)   **(P)**

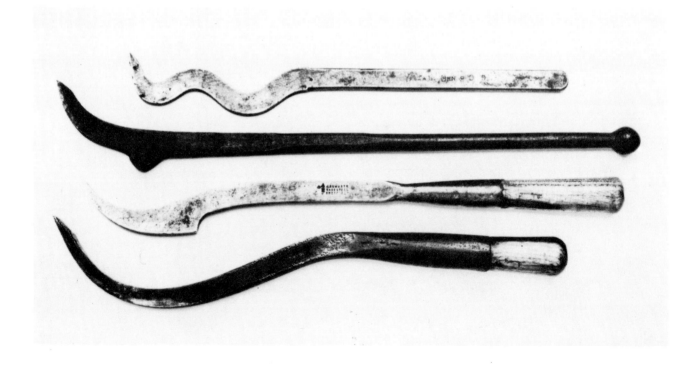

method was to sell the blades only, leaving the user to fit his own handle – which he might buy ready-made or make himself. For socket chisels the handle was simply a short, turned wood stub with a long tapering tail fitting into the socket, while home-made handles for tanged blades were usually a simple octagon, tapering towards the

144   **S. TYZACK & SON**

CARVING TOOLS
Sets of Ladies' Carving Tools

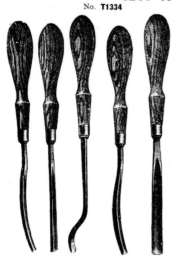

No. T1331

Number of Carving Tools in Set	6	12	18	24	36	tools.
No. **T1330.** Sets of Ladies' Carving Tools, Fancy Hardwood Handles, in Cardboard Boxes	10/6	19/-	28/6	39/-	58/6	set.
No. **T1331.** Sets of Ladies' Carving Tools, Hardwood Handled, in Pine Boxes ...	12/6	22/6	33/6	45/-	69/-	,,
No. **T1332.** Sets of Ladies' Carving Tools, Hardwood Handled, in Oak Boxes ...	15/9	26/6	39/6	52/6	78/-	,,
No. **T1333.** Sets of Ladies' Carving Tools, Hardwood Handled, in Mahogany Boxes	21/-	31/6	45/-	52/6	84/-	,,

SETS OF STRAW-COLOURED CARVING TOOLS

No. T1334   No. T1334

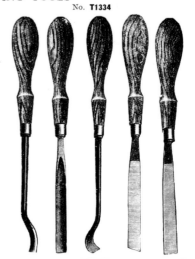

The construction of these Tools renders them specially well suited for ladies' use, although frequently adopted by workmen also. They are nicely finished, having straw-coloured tempered surface, and fitted with shaped Beech Handles.

No. **T1334**

Sets of	12	24	36	tools.
Price ...	19/-	37/6	46/-	per set.

HANDLED SPADE CARVING TOOLS

No. **T1335**   No. **T1335**

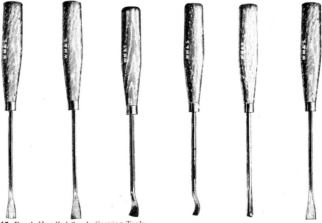

Set of **12**
Two sizes of each of
these shapes.

Set of **12**
Two sizes of each of
these shapes.

No. **T1335.** Set of 12 Beech Handled Spade Carving Tools   ...   .   **13/3** complete.

341, 343 & 345 Old Street, Shoreditch, London, E.C.

132
(*Left*) Carving chisels: this small selection of the vast range of carving chisels is from the Tyzack catalogue of *c*, 1920. The best known name in carving tools, Addis, was by then a mere brand-name of Ward and Payne.

133
Simple edge tools: four draw-knives, a Kent axe and three hurdle-maker's twybills. All the tools in this group are of Kentish origin: the twybill at top left was made at Sissinghurst, the axe at Tenterden and the adjacent twybills at Bethersden (*top*) and Charing.
(Christie's South Kensington)   (P)

blade. There was considerable variety in ready-made handles, as shown by the illustration from the Melhuish catalogue.

Gouges came in similar variety, with the extra permutation of 'in-cannel' or 'out-cannel', indicating that the sharpening bevel is on the inside or the outside of the curve respectively. Sash-pocket chisels are very thin, and were used for inserting in the saw-cut made at each end of the access slip in the pulley-stiles of a sash window; the cuts are about half-an-inch apart on the inside and outside, the short length between them being split by a sharp blow on the chisel, so forming a rebate. These are too specialized to be found vey often, and are sometimes mistaken for painters' stripping knives.

Draw-knives and spokeshaves

Of all the tools which have disappeared from most workshops today, the draw-knife is one that least seems to justify relegation to museum displays. It was particularly used by woodland and rural tradesmen, but is a superb tool for any work which requires the quick removal of more timber than can be quickly achieved with a plane but less than can be conveniently sawn off. The two-handed grip provided by the handles at each end of the blade gives far greater control than is possible with a chisel, and a draw-knife is particularly useful for forming chamfers, which can always be trimmed with a plane afterwards if a high finish is required. As with all edge-tools, early makers' stamps are the features to look for, but there are also different handles and different blades – some are straight, some curved

134
Coach shave: coachbuilders needed to work mouldings on curved surfaces, with changes in the direction of grain, which needed left- and right-hand cutters. This boxwood shave has a single cutter with two cutting edges.
(Christie's South Kensington)

edgewise, some curved on the flat side. The most extreme example of the latter is the cooper's inshave, which has a blade bent round almost into a ring in some cases, so that the handles are close together.

The spokeshave is a relation of the draw-knife, being effectively a draw-knife blade in miniature, fixed by two tangs in a wooden stock with a handle on each side. The blade is not easily adjusted, and some versions have the tangs projecting above the stock with a threaded portion held in place by wing-nuts. Because they are intended for use on curved surfaces, spokeshaves have very short soles, and wear is therefore rapid; some have a brass plate in front of the mouth to counteract this. The stock is usually of beech, but boxwood is also found quite often. The adjustable metal spokeshaves which came into the British Isles from America in the late nineteenth century, and were also made by Preston in Birmingham, eventually took the place of the wooden types. Many variations of the spokeshave were used by coach-builders, who worked almost entirely on curved surfaces and had shaves for forming grooves, rebates and mouldings. Most of these were home-made, although commercially produced examples are seen from time to time, and Preston even translated many of them into iron.

Axes and Adzes

Axes survive as timber-felling implements (although even there the portable chain-saw is fast displacing them) but many tradesmen used to use them for shaping timber, particularly shipwrights, wheel-wrights and coopers. Most such axes were side-axes – that is, unlike the felling axe, the blade has a flat side with the socket for the handle (haft) offset to the other side. This enables the tool to be used for trimming timber along the grain, usually in a vertical position so that the weight of the axe does much of the work. Many axes seem to survive from the eighteenth century or before: sometimes the steel cutting edge has been renewed. The rear part of the blade, including the haft-socket, was made of wrought iron, occasionally with impressed or even pierced decoration (see Chapter 4). The axe is one of man's oldest tools, and apart from the modern Canadian and Kent felling axes, there have been a surprising variety of forms made for

135
Axes, adzes and cooper's tools: 1925.

RICHARD MELHUISH LIMITED, LONDON, E.C.

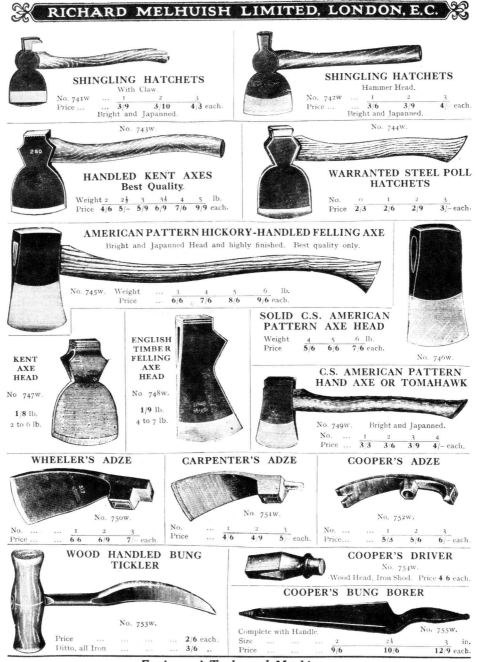

SHINGLING HATCHETS
With Claw.

No. 741W	1	2	3
Price	3/9	3/10	4/3 each.

Bright and Japanned.

SHINGLING HATCHETS
Hammer Head.

No. 742W	1	2	3
Price	3/6	3/9	4/- each.

Bright and Japanned.

No. 743W.

HANDLED KENT AXES
Best Quality.

Weight	2	2½	3	3½	4	5	lb.
Price	4/6	5/-	5/9	6/9	7/6	9/9	each.

No. 744W.

WARRANTED STEEL POLL HATCHETS

No.	0	1	2	3
Price	2/3	2/6	2/9	3/- each.

AMERICAN PATTERN HICKORY-HANDLED FELLING AXE
Bright and Japanned Head and highly finished. Best quality only.

No. 745W. Weight	3	4	5	6	lb.
Price	6/6	7/6	8/6	9/6 each.

KENT AXE HEAD

No 747W.

1/8 lb.

2 to 6 lb.

ENGLISH TIMBER FELLING AXE HEAD

No 748W.

1/9 lb.

4 to 7 lb.

SOLID C.S. AMERICAN PATTERN AXE HEAD

Weight	4	5	6	lb.
Price	5/6	6/6	7/6 each.

No. 746W.

C.S. AMERICAN PATTERN HAND AXE OR TOMAHAWK

No. 749W. Bright and Japanned.

No.	1	2	3	4
Price	3/3	3/6	3/9	4/- each.

WHEELER'S ADZE

No. 750W.

No.	1	2	3
Price	6/6	6/9	7/- each.

CARPENTER'S ADZE

No. 751W.

No.	1	2	3
Price	4/6	4/9	5/- each.

COOPER'S ADZE

No. 752W.

No.	1	2	3
Price	5/3	5/6	6/- each.

WOOD HANDLED BUNG TICKLER

No. 753W.

Price	...	...	...	...	2/6 each.
Ditto, all Iron	...	...	...	3/6 „

COOPER'S DRIVER
No. 754W.
Wood Head, Iron Shod. Price **4/6** each.

COOPER'S BUNG BORER

Complete with Handle.

No. 755W.

Size	...	...	...	2	2½	3	in.
Price	...	...	...	9/6	10/6	12/9 each.

Engineers' Tools and Machinery

136
More edge tools: top left is a French vineyard bill, next to it a French cooper's axe, and, in the centre, a French carpenter's twybill or besaiguë. The lowest axe in the picture is a Welsh miner's axe, and at the bottom left are two froes, much used in woodland trades for slitting long lengths of timber. (*Arnold & Walker sale*, 1979.)

different purposes and in different parts of the world, even within the last three centuries. There are not a great many collectors of axes, however, partly because not enough is known about them for individual examples to be easily identified and dated.

The adze is effectively an axe turned through ninety degrees, and was used for trimming timbers in the horizontal position, the user standing over the work with his legs astride and swinging the adze towards him. Apart from the normal long-hafted types held in both hands, there were small hand-adzes, some with short, steeply curved blades, known appropriately as bowl adzes. Although there is variety in the functional shape in adzes (and in the poll, as the stub on the opposite side of the socket from the blade is known) there tends to be less ornament than on axes.

One of the least familiar tools to the newcomer will be the twybill, whether in its English or French version. The former is a sort of woodworker's pick-axe, and it survived from medieval times into the present century as a hurdle-maker's tool. It was used in making mortices, and was obviously less precise than a mallet and chisel.

However, the French version, known as a *besaiguë*, was developed into a very long paring chisel, with a flat cutting edge at one end and a narrow one, like that of a mortice chisel, at the other. A short hand-grip of iron projected half-way along, and the tool was used in the standing position, with the upper end bearing against the user's shoulder. This, and the sheer weight of the tool itself, enabled the besaiguë to be used for cutting accurate mortices without the need for a mallet.

A point that should perhaps be made in the context of edge tools generally (including plane irons) is that, when the sharpening bevel is on one side only, a good edge can never be achieved once the flat side has been affected by rust. Where there is even slight pitting on that surface, nothing short of surface grinding is likely to make a decent cutting edge possible. This should be borne in mind if you like to use your tools rather than just keep them on a shelf.

Turnscrews

A turnscrew is a screwdriver by an old-fashioned name ('screwdriver' is thought to be of American origin). Which term you use does not really matter, but 'turnscrew' seems to me more appropriate for old-fashioned examples. Of the types shown in the Melhuish catalogue the 'Registered' is still with us, although plastic is often used for handles instead of wood, and the American type of fluted, parallel handle is also much favoured, especially by engineers and mechanics. I was pleasantly surprised, a year or two ago, to find London pattern screwdrivers on sale in a local ironmongers, but in general they are seldom seen now. As with chisels, early turnscrews tend to survive only in large examples, whose sheer size, and limited use, have preserved them. Many of these were home-made, or at least hand-forged by the local blacksmith, often from worn-out files.

Turnscrews earlier than the eighteenth century will be unlikely to appear, and when they do will probably not be woodworkers' tools, since wood screws only seem to have come into use in that century. Until well into the nineteenth century, their use was largely confined to fitting metal to wood (hinges, bolts and the like) in furniture and high-class architectural joinery; screws of this period have no points, and the slot in the head tends to be very narrow and slightly off-centre. Even the first machined screws had blunt ends. Pointed wood-screws appear to have come into general use in the 1850s. Some turnscrews were made as brace-bits, particularly the forked variety used for tightening the nuts on saw-handles. Ratchet and spiral-action screwdrivers were another American contribution of the late nineteenth century; North Bros.' 'Yankee' is the best known, and the name survives today as part of the Stanley range.

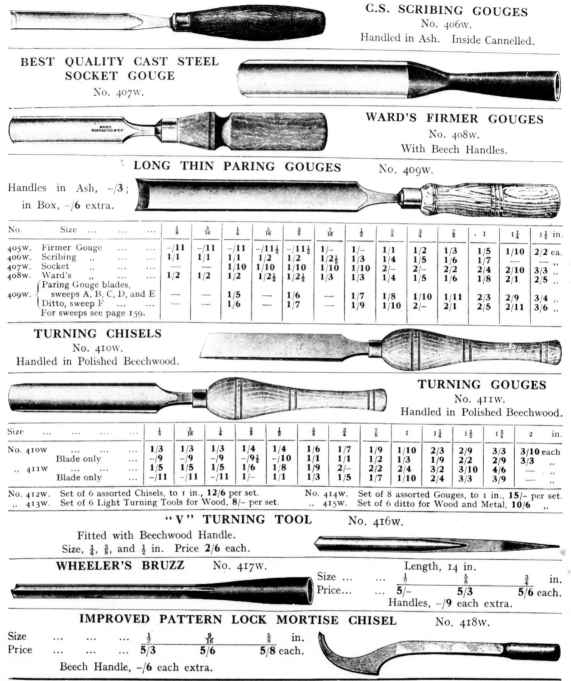

C.S. SCRIBING GOUGES
No. 406w.
Handled in Ash. Inside Cannelled.

BEST QUALITY CAST STEEL SOCKET GOUGE
No. 407w.

WARD'S FIRMER GOUGES
No. 408w.
With Beech Handles.

LONG THIN PARING GOUGES
No. 409w.

Handles in Ash, -/3;
in Box, -/6 extra.

No.	Size	...	...	$\frac{1}{8}$	$\frac{3}{16}$	$\frac{1}{4}$	$\frac{5}{16}$	$\frac{3}{8}$	$\frac{7}{16}$	$\frac{1}{2}$	$\frac{5}{8}$	$\frac{3}{4}$	$\frac{7}{8}$	I	$1\frac{1}{4}$	$1\frac{1}{2}$ in.
405w.	Firmer Gouge	...	...	-/11	-/11	-/11	-/11½	-/11½	1/-	1/-	1/1	1/2	1/3	1/5	1/10	2/2 ea.
406w.	Scribing „	...	...	1/1	1/1	1/1	1/2	1/2	1/2½	1/3	1/4	1/5	1/6	1/7	1/10	2/2 „
407w.	Socket „	...	...	—	—	1/10	1/10	1/10	1/10	1/10	2/-	2/-	2/2	2/4	2/10	3/3 „
408w.	Ward's „	...	...	1/2	1/2	1/2	1/2½	1/2½	1/3	1/3	1/4	1/5	1/6	1/8	2/1	2/5 „
409w.	Paring Gouge blades, sweeps A, B, C, D, and E			—	—	1/5	—	1/6	—	1/7	1/8	1/10	1/11	2/3	2/9	3/4 „
	Ditto, sweep F	...	...	—	—	1/6	—	1/7	—	1/9	1/10	2/-	2/1	2/5	2/11	3/6 „

For sweeps see page 159.

TURNING CHISELS
No. 410w.
Handled in Polished Beechwood.

TURNING GOUGES
No. 411w.
Handled in Polished Beechwood.

Size	...	...	$\frac{1}{8}$	$\frac{3}{16}$	$\frac{1}{4}$	$\frac{3}{8}$	$\frac{1}{2}$	$\frac{5}{8}$	$\frac{3}{4}$	$\frac{7}{8}$	I	$1\frac{1}{4}$	$1\frac{1}{2}$	$1\frac{3}{4}$	2 in.
No. 410w	...	...	1/3	1/3	1/3	1/4	1/4	1/6	1/7	1/9	1/10	2/3	2/9	3/3	3/10 each
	Blade only	...	-/9	-/9	-/9	-/9½	-/10	1/1	1/1	1/4	1/9	2/2	2/9	3/3	„
„ 411w	...	...	1/5	1/5	1/5	1/6	1/8	1/9	2/-	2/2	2/4	3/2	3/10	4/6	— „
	Blade only	...	-/11	-/11	-/11	1/-	1/1	1/3	1/5	1/7	1/10	2/4	3/3	3/9	— „

No. 412w. Set of 6 assorted Chisels, to 1 in., 12/6 per set. No. 414w. Set of 8 assorted Gouges, to 1 in., 15/- per set.
„ 413w. Set of 6 Light Turning Tools for Wood, 8/- per set. „ 415w. Set of 6 ditto for Wood and Metal, 10/6 „

"V" TURNING TOOL
No. 416w.
Fitted with Beechwood Handle.
Size, ¼, ⅜, and ½ in. Price 2/6 each.

WHEELER'S BRUZZ
No. 417w.

Length, 14 in.

Size	...	...	$\frac{1}{2}$	$\frac{5}{8}$	$\frac{3}{4}$ in.
Price	...	...	5/-	5/3	5/6 each.

Handles, -/9 each extra.

IMPROVED PATTERN LOCK MORTISE CHISEL
No. 418w.

Size	...	...	...	$\frac{1}{2}$	$\frac{9}{16}$	$\frac{5}{8}$ in.
Price	...	...	...	5/3	5/6	5/8 each.

Beech Handle, -/6 each extra.

Electric and Wireless Engineers
34

Gouges, special-purpose chisels, and various
screwdrivers. (Melhuish, 1925).

TOOL AND MACHINE MERCHANTS

SCREWDRIVERS

LONDON PATTERN No. 410W.

Cast Steel Blades, spring temper, cut into Strong Brass Ferrules.

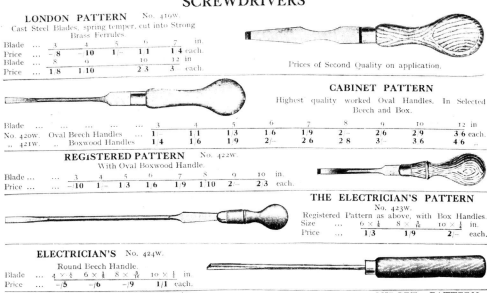

Blade	...	3	4	5	6	7	in.
Price	...	–/8	–/10	1/–	1/1	1/4	each.
Blade	...	8	9		10	12	in.
Price	...	1/8	1/10		2/3	3/–	each.

Prices of Second Quality on application.

CABINET PATTERN

Highest quality worked Oval Handles. In Selected Beech and Box.

Blade	...	...	...	...	3	4	5	6	7	8	9	10	12 in.
No. 420w. Oval Beech Handles	...				1/–	1/1	1/3	1/6	1/9	2/–	2/6	2/9	3/6 each.
,, 421w. ,, Boxwood Handles					1/4	1/6	1/9	2/–	2/6	2/8	3/–	3/6	4/6 ,,

REGISTERED PATTERN No. 422W.

With Oval Boxwood Handle.

Blade	...	...	3	4	5	6	7	8	9	10	in.
Price	...	...	–/10	1/–	1/3	1/6	1/9	1/10	2/–	2/3	each.

THE ELECTRICIAN'S PATTERN No. 423w.

Registered Pattern as above, with Box Handles.

Size	...	$6 \times \frac{1}{4}$	$8 \times \frac{3}{16}$	$10 \times \frac{1}{4}$ in.
Price	...	1/3	1/9	2/– each.

ELECTRICIAN'S No. 424w.

Round Beech Handle.

Blade	...	$4 \times \frac{1}{4}$	$6 \times \frac{1}{4}$	$8 \times \frac{3}{16}$	$10 \times \frac{1}{4}$ in.
Price	...	–/5	–/6	–/9	1/1 each.

POCKET SCREWDRIVER No. 425w. With 1½ in. Blade.

Handled in Beech	...	1/3 each.
,, Box	...	1/6 ,,

KNURLED BLADE No. 426w.

Blade	...	1½	2½ in
Price	...	–/9	–/10 each.

MIDGET PATTERN No. 427w.

Beech Handle and Brass Ferrule. Length of Blade, 1½ in.

Price	...	–/6 each.

POCKET SCREWDRIVER SET No. 428w.

This little Pocket Screwdriver Set, with three blades of assorted sizes, and one reamer for making or enlarging holes, will be found most convenient and attractive. In the explanatory cut the tool is shown in its closed form, with a side cut away to show the location of the different blades in the handle. The tool is fully polished, nickel-plated, and buffed. It is 3¼ in. long closed, and weighs 4 oz. Price 4 6 each.

PERFECT PATTERN SCREWDRIVER No. 429w.

Made from High-grade Sheffield Steel.

Length overall..	4½	6	8	10	12	14	16 in.
Price	...	–/10	1/–	1/4	1/10	2/6	3/3	3/9 ea.

PISTOL GRIP RATCHET SCREWDRIVER No. 430w.

This improved Screwdriver gives a firm grip to turn the most stubborn screw. It is a natural grip and will not tire the hand. Easy running screws may be quickly turned without taking the hand from the handles with the knurled ferrule.

Two widths of blades are furnished to fit common size screws. The reamer is handy in starting screws and similar work. The blades are changed without adjustments.

Set consists of handle, two blades, and reamer.

Length of Blade	...	...	...	...	...	...	6	8	in.
With two Blades and Reamer, price	...	...	...	...	...	...	9/6	10/– each.

Bronze Medal 1884—Gold Medal 1890

138
Turnscrew and grocer's hammer: large turnscrews like this (it is 28½ inches long) were often made by the local blacksmith out of old files; this one has the stamp of Thomas Beale, of Tenterden, Kent. The grocer's (or warehouse) hammer is a multi-purpose tool for opening or nailing up packing cases.

Saws

Saws, like chisels, consist of a blade and a handle, and therefore have a low survival rate from earlier than about the middle of the last century. Even frame saws, of which the most familiar is the bow or turning saw, seldom seem to be any earlier, although their frame is essentially non-wearing, the blade is replaceable and the type has a much longer history. The essential feature of the bow saw is its narrow blade, held in tension by a tourniquet at the top of the wood frame. The bandsaw and the lack of decorative curved features in modern work have between them made the bow saw a rare sight in today's toolshops, but it is still in use, and its form has changed little over many centuries. Larger versions were often used by woodcutters up to the last war, often with an iron rod and turnbuckle in place of the twisted string, but these have given way to tubular steel saws to which the term 'bow' is even more appropriate. Turning saws are often home-made, and very elegantly at that, but makers' names are sometimes found, usually stamped rather faintly on the frame, which is most frequently of beechwood, although other woods are used, including box.

The pad or keyhole saw also has a replaceable blade. It is perhaps too simple in design to offer great scope to the collector, but the

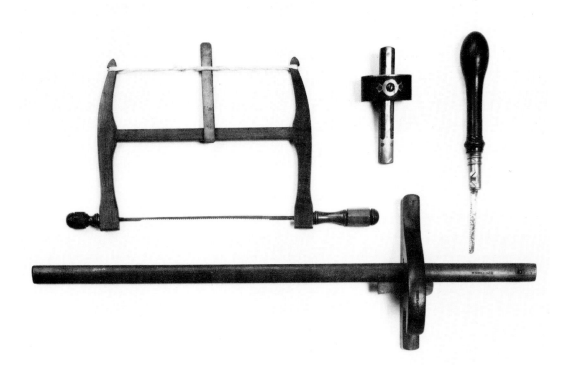

139
Saws and markers: the bow-saw is a 'gent's' model, with the blade only 10 in. long. The panel gauge at the bottom of the picture is mahogany, and the pad-saw handle is ebony. The ebony and brass mortice gauge was found by the author in a box which the previous owner of his house apologized for leaving behind.

turned wood handle is often of ebony, rosewood or box with a nicely detailed brass ferule. The handle has a slot right through its length, so that the blade can be adjusted to project no more than is necessary for the work in hand; this is necessary, as the thin blade is very easily bent. The pad saw is used for sawing enclosed cuts (a keyhole is a typical example) where an ordinary handsaw would be too wide and a bow saw would have no access. A larger version is the compass saw, which has an open pistol-grip handle; this has not been used much in the last few decades, although its design remains in a modified form with two or three interchangeable blades of different sizes. The latter type is intended mainly as a multi-purpose household tool rather than a workshop implement.

The saw most in use in British and American workshops is the handsaw, which has a tapered blade attached at the wider end to a closed handle. Only in the last twenty years or so has the handle lost some of its traditional decorative outline (particularly with the advent of plastic) and some wood handles, even in the present century, had carved surface decoration in addition to their well shaped outline and decorative brass escutcheons around the screws holding the handle to the blade. Another feature which survived well into the present century on traditional straight-backed saws was a cut-out section at

S. TYZACK & SON

1

SAWS

Our Hand Saws are ground about four gauges thinner at the back than at the front to ensure a perfectly easy clearance.

No. **T1a**

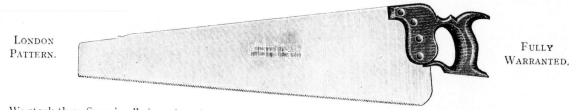

LONDON PATTERN.

FULLY WARRANTED.

We stock these Saws in all sizes of teeth, and send the usual when number of teeth to the inch is not specified.

Sizes	...	...	...	...	...	10	12	14	16	18	20	22	24	26	28 in.
No. **T1**.	Price, Best Quality				...	6/-	6/6	6/9	7/-	7/3	7/6	8/-	9/-	9/6	10/6 each.
,, **T2**.	,, Cheap Quality				...	1/9	2/-	2/3	2/6	2/9	3/-	3/6	3/9	4/2	5/- ,,
,, **T1a**.	Special Silver Steel				...	7/6	8/-	8/6	9/-	9/6	10/-	10/6	11/6	12/-	14/- ,,

SKEW BACK SAWS

Apple Wood Handle, American Pattern, Blades ground about four gauges thinner at the back than front. General finish and quality unequalled. English make.

No. **T3**

	No. T3			
Length	16	18	20	22 in.
Price	**7/3**	**7/6**	**7/9**	**8/6** each.
Length	24	26	28 in.	
Price	**9/6**	**10/-**	**11/-** each.	

the leading end of the back, with a small 'nib' at its rear end. The original purpose of this is not clear.

The term 'straight-back' is used to distinguish the traditional shape from the 'skew-back' developed in America in the 1870s by Henry Disston. Disston saws have long been a byword among wood-workers for their excellence, and it is interesting that, of all the handsaws that go through the tool sales at Christie's South Kensington, perhaps fifty per cent are Disstons, if not more. There is a steady demand for them from users, who consider pre-1939 Disston saws to be beyond compare.

For British collectors, unless they are specializing in saws, there is perhaps more interest to be found in saws by traditional Sheffield makers such as Spear and Jackson, Drabble and Sanderson, R. Groves and Sons, Taylor Bros, Hill late Howell and others. Names of major dealers such as Buck and Melhuish are also found on saws, although it is unlikely that they were the actual manufacturers of saws bearing

140
Handsaws, traditional and skew-back patterns. (Tyzack, *c.* 1920).

141
Back, bow, compass and pad saws (*c.* 1920).

2

S. TYZACK & SON

No. **T6**

DOVETAIL SAWS

		8	9	10 in.
No. **T6**.	Best Cast Steel, with Iron Back ...	**5 9**	**5 9**	**5 9** each.
,, **T7**.	Lighter Quality, ,, ,, ...	**4/-**	**4 3**	**4 3** ,,
,, **T8**.	Best, with Heavy Brass Back ...	**8/-**	**8/-**	**8/-** ,,
,, **T9**.	Ditto, Silver Steel	**10 6**	**10 6**	**10 6** ,,

No. **T10**

TENON SAWS

		12	14	16	18 in.
No. **T10**.	Cast Steel, with Iron Back	**6 3**	**6 6**	**7 6**	**10/-** ca.
,, **T11**.	Lighter Quality, ,,	**4 9**	**5 3**	**6/-**	**—** ,,
,, **T12**.	Cast Steel, with Heavy Brass Back ...	**9/-**	**10/-**	**11 6**	**14/-** ,,
,, **T13**.	Ditto, Silver Steel	**11 6**	**12 6**	**14/-**	**16/-** ,,

STEEL AND BRASS BACK SAWS. "Hy. Disston & Son's" Make

No. **T14**

Apple Wood Handles, with Polished Edges.

		8	10	12	14	16 in.
No. **T14**.	With Steel Blued Back	**8 6**	**9 5**	**10/-**	**11 6**	**13 6** each.
,, **T15**.	With Brass Back	**12 3**	**12 6**	**14/-**	**14 9**	**18 3** ,,

BOW SAWS AND FRAMES

No. **T16**

		8	10	12	14	16 in.
No. **T16**.	Beech Handled ...	**6 3**	**6 6**	**6 9**	**7/-**	**7 3** each.
,, **T17**.	Boxwood Handled ...	**7/-**	**7/-**	**8 6**	**9 6**	**10 6** ,,
,, **T18**.	Octagon Box Handle ...	**8 9**	**8 9**	**10/-**	**11/-**	**12 6** ,,
,, **T19**.	Blades for same ...	**–/3**	**–/3**	**–/3**	**–/4**	**–/4½** ,,

HARDWOOD HANDLED LIGHT BRASS BACK SAWS

No. **T20**

		4	5	6	7	8 in.
No. **T20**.	For Wood ...	**–/10**	**1/-**	**1/1**	**1/2**	**1 4** each.
	,, Metal ...	**–/9**	**–/10**	**–/11**	**1/3**	**1 9** ,,

No. **T22**

COMPASS SAWS

		12	14	16	18 in.
No. **T22**	... ...	**1 6**	**1 9**	**2/-**	**3/-** each.

PAD SAW HANDLES

No. **T23**

No. **T23**.	Pad Saw Handles, Beechwood, small	...	**1/-** each.
,, **T24**.	Ditto, ditto, large	...	**1 6** ,,
,, **T25**.	Ditto, Rosewood	...	**2 3** ,,
,, **T26**.	Ditto, Best Boxwood (small, **1 3**) ...	...	**2/-** ,,
,, **T27**.	Ditto, Best Ebony	...	**3/-** ,,

No. **T28**

PAD SAW BLADES

No. **T28**. To 12 in., **–/3** ; 13 & 14 in., **–/4** each.

341, 343 & 345 Old Street, Shoreditch, London, E.C.

their names. Most Disstons, and other saws of twentieth-century origin, have the name etched in an elaborate design on the blade, which is often difficult to decipher if rust has gained a foothold, but the simple, deeply stamped names on earlier Sheffield saws are easy to read.

Many of these early saws, if properly sharpened and set, are at least the equal of their modern counterparts in use. The three main categories are rip, crosscut and panel, these definitions referring both to the number of teeth, or points, per inch and to the shape of the teeth. Cross-cut saws have more or less equilateral triangle-shaped teeth, sharpened at an angle across the blade so as to produce a series of small knives, while rip teeth are nearly right-angled triangles in profile, sharpened straight across so as to produce a series of chisel edges for cutting along the grain of the timber. All saw teeth have a 'set' – that is, each tooth is bent slightly outwards, alternately to the left and right, so that the saw cut is wider than the blade, to prevent the latter binding in the cut. Numerous devices appeared in the late nineteenth and early twentieth centuries to simplify the setting of saw teeth, the majority of them being based on a pair of pliers with a miniature punch and anvil, adjustable to give exactly the right amount of set. The traditional method, requiring considerable skill, involved the use either of a special hammer and anvil or a saw 'wrest' or set consisting of a handled blade with a series of graduated slots in each side, in which the teeth were gripped and bent.

Backsaws, which have a steel or brass stiffening rib along the top of

142
Squares: mitre square (*bottom*), four sizes and styles of try-square and a sliding bevel (*top left*). Stocks are in ebony except for the small square at bottom left (mahogany) and the sliding bevel, which is ebony on one side (*uppermost in the picture*) and pale rosewood on the other.

the parallel blade, are used for precision cuts in joinery, and range from small 'gents' or bead saws of only six inches or so in length up to tenon saws of fourteen or sixteen inches. Larger sizes are occasionally found, but are usually out of special mitre-cutting devices. The term 'gents', incidentally, was at one time used of many small tools considered suitable for use by gentleman amateurs; some were even designated 'ladies' (as in the sets of carving tools shown in the Melhuish catalogue, which dates from as late as 1925).

Frame saws include the turning and log-cutting saws already mentioned, but larger, sturdier versions have long been used, particularly on the Continent, for general sawing in preference to handsaws. Even the initial sawing up of tree-trunks into planks was often done with a frame saw operated by two men, although British sawyers tended to favour an unframed saw with a long handle ('tiller') at the top and a removable handle ('box'), usually of turned wood, at the lower end. These saws are normally tapered, with the teeth in a straight line, unlike the two-man cross-cut saw, which has the teeth on a curve, making the blade wider in the middle than at each end.

Marking and Measuring

'Mark twice, cut once' is an old adage in woodworking classes, and the beauty of many old marking instruments should be encouragement for anyone to obey the instruction with pleasure. The need for precision necessitated good quality materials and workmanship, and the woods found are predominantly boxwood, ebony and rosewood, set off by brass fittings and reinforcements.

Rules were made in boxwood from the eighteenth century, if not before, to the present day. Some early types found in joiners' shops incorporated brass slides for calculating prices and quantities of timber. The arch-joint typical of English rules is a masterpiece of precision, enabling the rule to unfold into an accurate straight line in both planes. The continental type of 'zig-zag' rule, in which each section is simply pivoted on top of the next, never seems to have found favour in Great Britain, even for rough carpentry.

For marking lines at right-angles to the edge of the timber, a try-square is used; most commonly, this has a thin 'blade' fixed into a thicker wood stock. The blade on British examples is usually of steel but all-wood versions are common on the Continent. Another all-wood form is similar to the draughtsman's set-square, a piece of flat wood cut to a right angle, the 'hypotenuse' of the triangle sometimes being shaped as a French curve. The steel blades on British squares are held in the stock by rivets in brass escutcheons, which are found in a variety of plain and decorative shapes. The inside edge of the stock is normally faced with brass, and occasionally the outer edge is also. The wood is usually either rosewood or ebony, and the various escutcheon designs offer plenty of scope for collecting.

S. TYZACK & SON

MARKING AND CUTTING GAUGES

MARKING

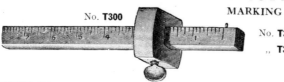

No. **T300**

No. **T300**.	Beechwood Gauge, Rule Marked ...	–/**8** each.
,, **T301**.	Ditto, Polished, with Adjusting Steel Point	**1 3** ,,

No. **T303**

No. **T302**.	Best Beech Marking Gauges	**1/–** each.
,, **T303**.	Ditto, Brass Plated on Face	**1/6** ,,
,, **T304**.	Ditto, Head Faced with Brass	**2/6** ,,
,, **T305**.	Ditto, London Pattern, Oval Head ...	**2/3** ,,

CUTTING

No. **T306**

No. **T306**.	Best Beechwood Cutting Gauges	**1/3** each.
,, **T307**.	Ditto, Brass Plated	**1/10** ,,
,, **T308**.	Best Hardwood Cutting Gauges ...	**2 6** ,,
,, **T309**.	Ditto, Brass Hooped	**3 3** ,,
,, **T310**.	Head Faced with Brass and Brass Hooped	**3 9** ,,

No. **T311**

No. **T311**.	Patent Wedge Beech Marking Gauges, small size	–/**9** each.
,, **T312**.	Ditto, with Plated Head	**1/3** ,,
,, **T313**.	Ditto, with Brass Faced Head	**2/–** ,,

NEW PATTERN EBONY CUTTING GAUGE

No. **T314**

No. **T314**. Price **5/–** each.

IMPROVED OVAL HEAD CUTTING GAUGE

No. **T315**

No. **T315**. Price **7/3** each.

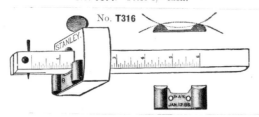

No. **T316**

No. **T316**. Polished Beechwood, Rule Marked Stem, Adjusting Steel Point, with Two Brass Ribs or Projections, which ensure steadiness when running a gauge line around curves of any degree ... ... **1/4** each.

No. **T317**

No. **T317**. Rosewood, Brass Adjustable Slide, Brass Screw, Faceplate ... **5 9** each.

No. **T318**

No. **T318**. Solid Brass Head, Ebony Thumb-screw End, Mortise Gauges ... **11/–** each.

341, 343 & 345 Old Street, Shoreditch, London, E.C.

143
Gauges: a selection of the many patterns available.
(Tyzack, c. 1920).

For marking angles, there is the mitre square, similar to a try-square but set at forty-five degrees, the mitre templet, cast in brass, iron or aluminium, the sliding bevel which is an adjustable version of a try-square, and the boatbuilder's bevel, which differs from the sliding variety in having two blades of unequal length, which are pivoted but not sliding. Both kinds have slots in the stocks into which the blades can be folded when not in use.

Marking gauges are used for marking lines parallel with the edge of the timber, and have a sliding stock or fence on a stem with a point mounted near one end. The stock is held in position by a captive wedge or thumbscrew, or, on some versions, is threaded to screw along a threaded stem, and has a second section to screw up against it as a lock-nut. Beech is the wood most often used for marking gauges, but the more elaborate mortice gauges are frequently found in ebony or rosewood with brass fittings – on some, the stem itself is of brass, and one pattern particularly popular with collectors has a flanged brass stock on an ebony stem. Mortice gauges have two points, one of which is independently adjustable. Usually this is attached to a brass slide, which may have a screw or thumbscrew adjustment, although another type has two stems with a point in each. Cutting gauges are similar to marking gauges except that the point is replaced by a removable knife-edge blade. Attractive ebony and brass versions are sometimes seen. All-iron marking and mortice gauges were developed in America, but this is one instance in which the traditional tool has seen off the newcomer and remains the more popular type.

Many cabinet-makers made their own gauges, particularly in the case of panel gauges, which have stems usually between about twenty and thirty inches long. The fence on these is rebated to fit over the edge of the timber and keep the stem at a constant height, and has a curved top usually shaped to form an elegant hand-grip. Mahogany is often used, a characteristic of many user-made tools.

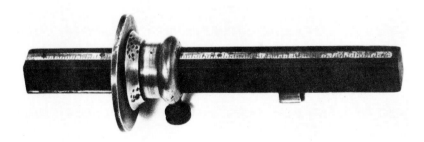

144
Mortice gauge: many mortice gauges are beautifully made in ebony and brass, but exist in too great numbers to excite collectors. This pattern, with a brass head on an ebony stem, is relatively uncommon and can bring £40–£50 at auction.
(Christie's South Kensington)

145
Circle cutter: used for cutting large holes (for example, in washstand tops), with a pre-drilled hole for the pin in the centre. This example is in dark Cuban mahogany.
(Christie's South Kensington)   (P)

Compasses and callipers are used less in woodworking than they once were, but these are essentially non-wearing tools and there is considerable variety among early examples extant. As with other all-metal tools, dating is not always easy, although some of the more attractive examples have inscribed dates. Compasses are used for marking and measuring distances between two points, and for describing arcs and circles, while callipers measure inside or outside cylindrical objects. Construction is basically similar in either case. A particularly attractive form of compass is the French cooper's wooden bow compass, in which the two arms are adjusted by a cross-piece threaded at each end with opposite-handed threads. This is thought to be a survival of a type once in common use in various trades.

146
Brass-mounted spirit level: made by Edward Preston and Sons of Birmingham, as well known for their rules and levels as for their planes.
(Christie's South Kensington)   (P)

147
Cast iron at its most delicate; an adjustable spirit level by the Davis Tool and Level Co., patented by L. L. Davis in the USA in 1867. This one is 18 in. long.
(Christie's South Kensington)   (P)

148
Callipers and compasses: a particularly fine late
17th-century pair of English callipers, inscribed
with the maker's name (see detail) and that of the
owner – Benjamin Cooper, of Codnam, Suffolk. The
compasses are of a type often found in France.
(*Old Woodworking Tools*)

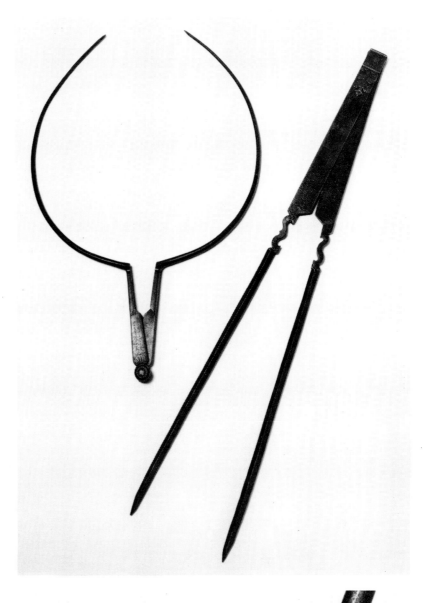

149
Detail of the inscription on the callipers.
(*Old Woodworking Tools*)

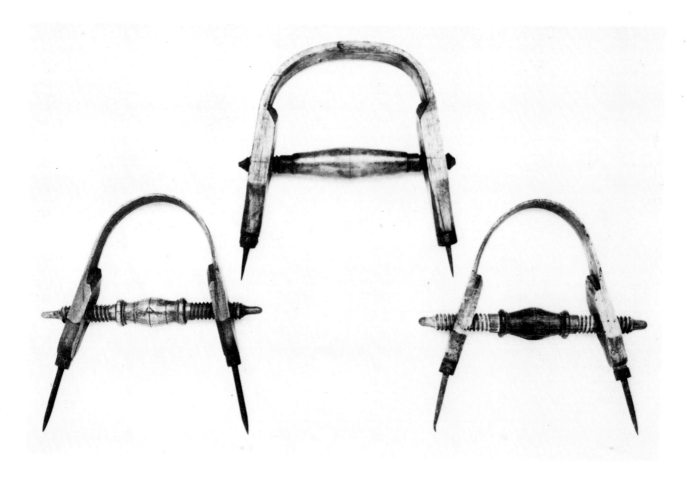

Hammers

One of the most commonplace and basic of all tools, the hammer is also one of the more difficult to collect. Even though a hammer does not wear out to the extent that a cutting tool does, early examples seldom appear on the market, and when they do are likely, yet again, to be of a specialized nature such as the very splendid example illustrated on page 00. Probably one reason why so few tool-kits seem to contain hammers is that, when the original owner ceased to ply his trade, his hammer was taken into general household use by himself or his family, and never replaced in the tool-chest. Woodworking hammers fall broadly into two categories; those with a flat pane (pein) opposite the normal circular striking face, and those with a claw for withdrawing nails. The latter type used often to be forged with plates extending up the handle (this style is known as a 'Canterbury' hammer) but present-day claw hammers usually have an 'adze-eye' — that is, the handle fits into an extended rectangular socket which provides a more secure hold than if only the thickness of the head

150
Wooden screw compasses: three examples from the Arnold and Walker sale. In the 17th and 18th centuries, these were in common use by coopers throughout Europe, but latterly became mainly confined to France.
(Christie's South Kensington)   (P)

151
Hammers and a cooper's shave: 1925.

RICHARD MELHUISH LIMITED, LONDON, E.C.

HAMMERS

BEST C.S. JOINERS OR WARRINGTON PATTERN

No. 712W.

No. ...	0	1	2	3	4	5	6	7	8	
Price	11	1	1·1	1·3	1·4	1·6	1·8	1·9	2·2	each.

C.S. EXETER OR LONDON PATTERN

No. 713W.

No. ...	0	1	2	3	4	5	6	7	8	
Price	1/2	1/3	1/4	1/5	1/6	1/8	1/9	1/11	2/1	each.

PIN HAMMERS

CROSS PANE   No. 714W.

Weight, 2 oz.	Price...	...	...	...	1/2 each.
4 ,,	,,	...	...	...	1/3 ,,

BALL PANE   No. 715W.

Weight, 2 oz.	Price	...	...	...	1/2 each.
4 ,,	,,	...	...	...	1/3 ,,

BEST CANTERBURY HAMMER
With Polished Oval Handle.

No. 710W.

No. ...	0	1	2	3	4	5	6	
Price ...	2/6	3/-	3/3	3/6	3/9	4/-	4/6	each.

IMPROVED SOLID C.S. FARRIER'S HAMMER

No. 717W.

Price ... ... 3/6 each.

Weight of head, 8 oz.

ADZE EYE CLAW   No. 718W.
These are of Solid Steel and highly finished.

No.	0	1	1½	2	3	
Weight of head	28	20	16	13	7	oz.
Price	4/6	4/3	4/-	3/9	3/6	each.

CAST STEEL SHOEMAKER'S
With Polished Handle.

No. 719W.

No.	0	1	2	
Price...	1/9	1/10	2/3	each.

GEOLOGISTS' HAMMERS

With Oval Ash Handles.   No. 720W.

Price ...	4	8	12	16	oz.
	2/-	2/6	3/-	4/-	each.

With Oval Ash Handles.   No. 721W.

Price ...	4	8	12	16	oz.
	2/-	2/6	3/-	4/-	each.

CLAW TACK HAMMER

No. 722W.

Black.

Price ... ... 1/- each.

WARRINGTON TACK HAMMER

No. 723W.

Black.

Price ... ... 1/2 each.

WROUGHT COAL HAMMER

No. 724W.

With Steel Face and Points.

Weight, 1½ lb. Price ... ... 3/3 each.

COOPER'S SHAVE

No. 725W.

Price 3/-

HAMMER PINCERS

No. 726W.

Length, 6 in. Price 3/6 each.

Electric and Wireless Engineers

were used. Once again, this appears to have been an American development, and many continental hammers still have straps, which differ from the English variety in being separate from the head, passing through the eye alongside the handle and bent outwards at the ends to prevent the head from literally flying off the handle. The handle itself is traditionally made of ash, a very resilient timber which is still used for lighter hammers such as the joiner's Warrington pattern, but the heavier claw hammers are now normally handled in hickory.

With the exception of the marking and measuring instruments, most of the tools mentioned in this chapter have tended not to receive as much attention from collectors as they perhaps deserve. It is understandable that a piece of elaborate workmanship like an Ultimatum brace should bring a higher price than a humble Canterbury hammer, but it does seem that the latter is the rarer tool, and it may be that the time has come for collectors to extend their sights and thinking laterally, if not downwards!

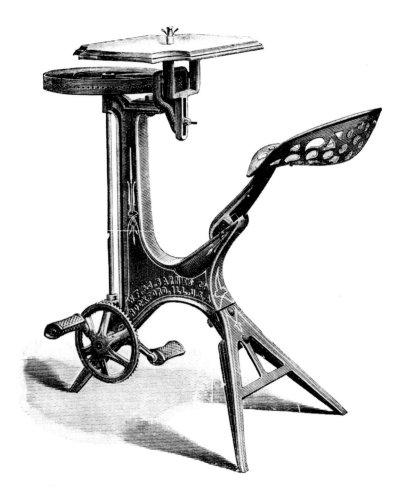

152
Spindle moulder: most of the machines which seem to owe their existence to electrical or other power sources were adapted to foot power in the 19th and early 20th centuries. This American machine appeared in a Tyzack catalogue in the early 1920s.

GLOSSARY

arris   Sharp right-angled corner (a brick, for example, has 12 arrises)

boxed   (of wood planes) With boxwood insert(s) in the sole to resist wear

chuck   That part of a brace which grips the bit

cutter   The blade of a plane, especially a metal plane

felloe   Curved section forming part of a wheel rim

fence   Section of a moulding, grooving or fillister plane projecting below the level of the sole at one side, serving to guide the plane parallel to the edge of the timber being worked

frog   (in Bailey-type planes) Movable section forming a bed for the cutter

infill   Wood used to fill the space between the plates of a metal plane or framed brace

iron   The blade of a plane

mouth   The slit in the sole of a plane through which the cutting iron protrudes

nicker   Small blade in front of the main cutter (e.g. on a fillister) which makes an incision at the edge of the cut

pad   Early name for a chuck, often consisting of a separate wood shank permanently fixed to the bit and removable from the brace

poll   Hammer-like projection on an adze or axe, on the opposite side of the haft-socket to the blade.

prow   Projecting section at the front of a skate

quirk   Returned section of a convex moulding forming a crevice – strictly, with a small flat at the bottom (see ogee diagram, p. 39)

rails   In a frame (e.g. a door or a window sash) the main horizontal members

seat   (in Bailey-type planes) The machined surface to which the frog (q.v.) is fixed

shooting   Process of planing an edge true – in the case of short edges (e.g. mitre joints) normally done on a shooting board

shot   (of wood planes) Shortened by planing off an end on a shooting board

skate   Narrow steel sole on a plough or grooving plane

sole   (in British usage) The bottom, or wearing surface, of a plane

spur   See 'nicker' – 'spur' is normally used of the type found in American combination planes

stiles   In a frame (e.g. a door or a window sash) the two outer upright members

stock   The main body of a plane or other tool

stop   (as in depth stop) That part of a moulding, grooving or fillister plane which controls the depth of cut

stuck mouldings   Mouldings worked directly on a work piece, as distinct from those 'applied' as separate pieces

throat   The opening above the mouth through which the shavings escape

web   That part of a brace (or crank of any kind) connecting the handle to the central axis

PRICE GUIDE

Objects shown in the illustrations listed below were sold at Christie's South Kensington on the dates and at the prices indicated. The dollar conversions have been made at the rate of the dollar on the date of the sale to the nearest dollar. When several objects are shown in the same illustration the prices are listed in the order in which the objects are listed in the captions.

Illus.	Sale Date	Price £	$
4	21.10.80	250	607
11	23.4.79	280	580
12	16.12.82	170 (mitre & stair-rail)	274
		30 (strike-block)	48
15	17.9.81	150	277
		140	259
		45	83
16	17.9.81	30	55
		420	777
		360	666
17	17.9.81	95	176
18	5.7.79	35	78
23	14.5.81	300	621
24	17.9.81	1100	2035
25	20.11.80	210	499
		140	333
37	14.8.80	320	758
38	26.2.81	170	377
		150	333
		440	977
		600	1332
		85	189
		28	62
42	21.10.80	75	182
43	17.9.81	420	777
49	24.4.80	200	452
51	14.2.80	200	462
53	21.10.80	75	182

Illus.	Sale Date	Price £	$
54	16.12.82	38 (with other items)	61
		170	274
		420	676
		55	89
		25	40
55	5.7.79	130	290
56	16.12.82	140 (with other items)	225
		130	209
58	20.11.80	70	166
59	23.4.79	240	497
61	21.10.80	130	316
64	21.10.80	130	316
66	16.12.82	200	322
75	16.12.82	30	48
82	2.9.82	550	946
107	16.12.82	130	209
		35	56
108	20.11.80	170	403
		70	166
		320	758
		140	332
		160	379
		700	1659
110	20.11.80	1500	3555
112	24.4.80	580	1311
118	17.9.81	150	277
122	14.2.80	150	346
123	16.12.82	60 (turning plane only)	97
124	26.2.81	400	888
130	23.4.79	top £62 (with one other)	128
		centre two £55 together	144
		bottom £38 (with one other)	79
133	21.10.80	42	102
		20	49
		18	44
		48	117
		65	158
145	17.9.81	60	111
	21.10.80	75	182
146	21.10.80	65	158
147	22.11.79	48	104
150	23.4.79	600	1242
		320	662
		95	197

BIBLIOGRAPHY

General

BERNT, WALTHER, *Altes Werkzeug*, Munich, 1939, reprinted 1977. Black and white photography with German text of medieval tools in German museums and collections, giving current locations of those tools.

DUNBAR, J. M., *Antique Woodworking Tools*, New York, 1977.

FELLER and TOURET, *L'Outil*, Belgium, 1969. 113 colour illustrations and French text of the superb tools in the Troyes Museum. Excellent drawings of details illustrating different period styles.

GOODMAN, W. L., *The History of Woodworking Tools*, London, 1964. Over 200 illustrations, line drawings and photographs. This is *the* current history and treats each kind of tool separately.

LAYTON, DUDLEY A., *Let's Collect Old Woodworking Tools*, Norwich, 1977.

MERCER, HENRY C., *Ancient Carpenters' Tools*, New York, 1975. Now in its third edition, this book was published in 1929 and written by a pioneer collector in the USA. It contains some 250 illustrations, mainly photographs, identifying and illustrating hand tools from the earliest times to the twentieth century.

ROBERTS, KENNETH D., *Some 19th-Century English Woodworking Tools*, New Hampshire, 1980. A documented history with many illustrations of edge tools, braces, and other joiners' tools manufactured in Sheffield and Birmingham. Much detail of makers and marks.

SALAMAN, R. A., *Dictionary of Tools Used in the Woodworking and Allied Trades 1700–1900*, London, 1975. The most comprehensive reference work, containing some 2600 entries and 2000 illustrations of traditional hand tools.

VELTIER, A. and LAMOTHE, M.-J., *Le Livre de L'Outil*. A sumptuous collection of 250 illustrations in full colour, and many in black and white, organized around eleven woodworking and thirteen other trades. Eighteenth-century engravings are mixed with photographs of tradesmen using the tools. The most beautifully illustrated book on tools available. French text.

WALKER, PHILIP, *Woodworking Tools* (Shire Album), Aylesbury, 1980.

Planes

GOODMAN, W. L., *British Planemakers from 1700*, 2nd edition, Needham Market, 1978. Contains nearly 900 makers of wooden and metal planes from the earliest known English maker to the middle of this century. Gives details of marks, addresses, and history. Invaluable for identifying makers and determining the age of planes.

ROBERTS, KENNETH D., *Wooden Planes in 19th-century America*, 2nd edition, New York, 1982. A full survey of both American and Canadian planes with checklists. Discussion of English planes is made where relevant.

SELLENS, ALVIN, *The Stanley Plane*, Kansas, 1975. A history and descriptive survey of all Stanley planes from the beginning of the Company in the 1850s through to the 1970s. Fully documented and illustrated.

SELLENS, ALVIN, *Woodworking Planes*, New York, 1975. A descriptive register of the basic information on the many varieties of American wooden planes. Emphasis is on the identification, proper description, and proper name of each variety.

SMITH, ROGER K., *Patented Transitional and Metallic Planes in America 1827–1927*, Lancaster, Mass., 1981. An exhaustive study of the forerunners of the modern metal plane. Extremely well documented with 350 superb photographs, including 41 in colour. Many of these planes were exported to Great Britain and appear for sale here.

Reprints of Old and Antique Books and Catalogues

BERGERON, *Manuel du Tourneur*, Paris, 1981. Facsimile of the 2nd edition of 1818, containing 96 plates. 3 volumes.

DIDEROT, *The Complete Plates*, New York, 1969. A reprint in one volume of the 3132 engraved plates from the 12 volumes of the great *Encyclopédie* by Diderot and d'Alembert, which appeared between 1751 and 1780. This volume illustrates 200 trades and their tools, including all the possible woodworking trades.

FERON ET CIE, *A la Forge Royale* (Paris 1927), USA 1981. Many illustrations, most of the tools shown being of wood. Reprinted with English translations of the captions.

HOLTZAPFFEL, *Construction, Action and Application of Cutting Tools Used by Hand*, USA, 1981. First edition was in 1975: this is the most comprehensive treatise on hand-tools ever published. It has chapters on planes, boring tools, screwcutting tools, saws, files, shears, and punches and discusses the theoretical and geometric aspects of the tools.

Knight's American Mechanical Dictionary, USA 1979. A 4-volume work started in 1876 by a U.S. patent commissioner, giving thousands of examples of 19th-century inventions, many of them hand tools. There are some 5000 engravings.

MATHIÉSON, 1899, USA 1979. Eight edition of the catalogue of this important Glasgow manufacturer. Apart from the wide array of tools, this catalogue is very useful for identifying mouldings from the comprehensive profiles of single, double, and triple ironed planes.

MILLERS FALLS, 1915 catalogue, Lancaster, Mass., 1981.

NORRIS/SPIERS, New Hampshire, 1979. A collection of small catalogues and leaflets from both these famous metal-plane makers, bound together.

PLUMIER, *L'Art du Tourneur*, Paris, 1979. Facsimile of the first book on turning, containing 80 plates. (From the 1749 edition).

PRESTON, *1901 Catalogue*, New Hampshire, 1979. Contains a full selection of wood and metal tools and rules and levels from this Birmingham manufacturer (1815–1933). There is also a range of rules from the firm of Thos. Bradburn, acquired by Preston.

RABONE, 1892 Catalogue, New Hampshire, 1982. Probably the most extensive catalogue of rules published in the nineteenth century. There are 96 pp of rules and levels, including 28 in full colour.

ROBERTS, KENNETH D., *Tools for the Trades and Crafts*. Over 200 pp with 134 full-page reprints of tools and implements of the first forty years of the nineteenth century, taken from Birmingham pattern books of the 1840s.

Smith's Key to the Various Manufactories of Sheffield, Vermont, 1975. The most detailed of the early British tool catalogues. A beautifully printed pictorial inventory of over 1000 engravings of tools and cutlery produced in Sheffield in 1816, with a contemporary price list keyed to the text.

STANLEY RULE AND LEVEL CO, New Hampshire, 1979. Eleven catalogues: 1859, 1867, 1870, 1874, 1877, 1879, 1884, 1888, 1892, 1898, and 1909.

D. STOLP, Zutphen, Netherlands 1915, USA 1981. Typical Dutch tool catalogue.

STRELINGER, 1897 Catalogue, USA, 1979. Catalogue of a major American distributor of tools from all over the world. Detailed and interesting contemporary comparisons of these tools are interspersed throughout.

J. WEISS U. SOHN, Vienna, 1909 Catalogue, USA 1980. Shows great variety of Continental tools; goosewing axe, twybill, wooden bitstock. The only carving tools listed are Addis 'Best British Steel'.

Those books listed above which cannot be found in bookshops are available from Roy Arnold, Old Woodworking Tools and Books, 77 High Street, Needham Market, Suffolk.

INDEX